£9.99,

Ways of knowing

MANCHESTER
UNIVERSITY PRESS

To Jonathan and Edwin

Ways of knowing

A new history of
science, technology and medicine

JOHN V. PICKSTONE

Manchester University Press

Published by Manchester University Press
Oxford Road, Manchester M13 9NR, UK
www.manchesteruniversitypress.co.uk

British Library Cataloguing-in-Publication Data
A catalogue record for this book is available from the British Library

ISBN 0 7190 5993 3 *hardback*
       0 7190 5994 1 *paperback*

First published 2000

07 06 05 04 03 02 01 00   10 9 8 7 6 5 4 3 2 1

Typeset in Sabon
by Servis Filmsetting Ltd, Manchester, UK
Printed in Great Britain
by Biddles Ltd, Guildford and Kings Lynn

Contents

Acknowledgements

A book of this scope reflects and creates many debts. I began to study history of science in 1968, after initially training as a physiologist. I would like to thank all who contributed to my first education – in Burnley, Cambridge and Kingston Ontario; and all with whom I have studied and worked as a historian – in London (UCL and Chelsea), Minneapolis, and in Manchester (first in Donald Cardwell's department at UMIST, and from 1986 at Manchester University). I am especially grateful to all the past and present colleagues who have helped create our Centre for the History of Science, Technology and Medicine (CHSTM).

A book such as this would be impossible without the work of hundreds of scholars, and I count myself lucky to be part of an international community of uncommon creativity and collegiality. I have tried to be generous with references in the text – to suggest further reading as much as to substantiate my points – but those acknowledgements fall far short of the true debt. If I have forgotten to record your contribution, I ask for charity. If, worse still, I have misrepresented your views, or mangled a topic which you know well, please let me know at the address below and I will hope to make corrections. This is a novel book; I would welcome discussion.

Since I first began to formulate the leading ideas of this text, I have been fortunate to receive the questions and support of several good friends and colleagues. David Edgerton has been a stimulating critic throughout, Janet Carsten encouraged me in the early days, and Jim Secord arranged the 'Big Picture' conference at which some of the ideas were first presented to the British

Society for the History of Science. Jon Harwood, Mick Worboys and Jack Morrell have been wise counsellors, and Jenny Stanton contributed more than a crucial clarification. When I have worried that I was wrong about physical sciences, I have been variously reassured by Simon Schaffer, Crosbie Smith, Andy Warwick, Robert Bud, Joe Marsh, Anna Guagnini, and by my memories of Donald Cardwell and Wilfred Farrar. Gaby Porter taught me much about museums (and their effects). I am grateful to them all, and to Seona Owtram for her kindness and understanding.

I have presented related material at several seminars in Britain and continental Europe: in Manchester; for the Institute for Historical Research and the Science Museum in London; the Society of the History of Technology meeting at Uppsala; the INSERM-158 Unit and La Villette in Paris; the Science Dynamics Department in Amsterdam; the History of Medicine Institute at Göttingen University; and the Max Planck History of Science Institute in Berlin. The expansion of European links in history of science, technology and medicine has been one of the best features of the past decade, and I appreciate the continuing friendship of historians in those cities.

Nearer to home, I owe much to my colleagues and former colleagues in other departments at Manchester: to John Breuilly who convinced me that my work could be of use to political historians, to Tim Ingold for discussions on material culture, to Penny Harvey for the ESRC seminars on Technology as Skilled Practice, to Marcia Pointon for advice on art history and intellectual verve, and to Katharine Perera for the cheerful clarity which dissolves the difficulties we bring to Pro Vice-Chancellors. That university staff here and elsewhere in Britain maintain such collegiality in the face of ever-increasing bureaucracy says much for their spirit and intellectual commitment, and little for our ability to dissuade government from make-work schemes that waste millions of tax-pounds.

In preparing the book for press, I have drawn heavily on the kindness and tolerance of colleagues in CHSTM and elsewhere. Jon Harwood, Jon Agar, Jeff Hughes, Vladimir Jankovic, Paula

Guest, Lyuba Gurjeva, Carsten Timmermann, Gary Hausman, Abigail Woods and Sharon Mathews kindly read drafts and gave useful comments. Mick Worboys, Roy Porter, David Edgerton and Simon Schaffer were also generous and constructive critics. Joan Mottram, as ever, helped in the preparation of the text, and so did Yvonne Aspinall and Helen Valier. I thank them all; no one should blame them for the book's imperfections.

Vanessa Graham and Alison Whittle at Manchester University Press and copy-editor Carol Lucas were very good to deal with, and at Chicago University Press Susan Abrams gave the encouragement only possible from a publisher for whom good history of science is a cause as well as a business. I am most grateful to Damien Hirst for permission to use the cover illustration 'Forms without Life', and to Richard Wentworth for his ready advice.

The Wellcome Trust has been a generous supporter of our work in Manchester and kindly provided me with the partial sabbatical when I put together some of the drafts from which the book is built. I am deeply appreciative of all the Trust has done for history of medicine (and of much that it may still do).

Finally, I thank my families – especially my mother and all who care for her, and Vivienne for her companionship and support. This is a book about the past, but it is meant for the future, so it is dedicated to our sons Jonathan and Edwin.

John V. Pickstone
Wellcome Unit and Centre for the History of Science,
Technology and Medicine
University of Manchester
http://www.man.ac.uk/chstm/

A note to the reader

This book has a novel form and is intended for a wide range of readers. It outlines the histories of science, technology and medicine, *not* in a single chronological sequence, nor discipline by discipline, but as different *ways of knowing*, each with its own history. These ways of knowing were interconnected in various ways; they could all be found at any period, but they varied in relative importance over time. I have chosen to call them *world-readings* (or *hermeneutics*), *natural history*, *analysis*, *experimentalism* and *technoscience*, and to each I have devoted a chapter (or two). Within each of the main chapters there is *some* material from the Renaissance through the five centuries to the present day, but later chapters focus heavily on more recent times.

Thus the picture builds up through the core of the book: we see the layers of knowledge changing, overlapping and growing over centuries; we recognise the peculiar knowledge-formations characteristic of each period, but we can also appreciate the *variety* of 'knowings' in play at any given time. This approach seems realistic, open-ended and easy to follow – but it is substantially different from most such histories.

For that reason I have explained my method fully in Chapter 1, which is partly intended for readers who already know some history of science and want to discover how my scheme relates to others, or how my ways of knowing are supposed to relate to each other and to ways of making (or technology). If these are not *your* primary concerns and you find the methodology a little 'heavy', then just skim the first chapter and read from Chapter 2.

This second chapter again stresses the variety of ways of

knowing, but it focuses on world readings – the shifting 'meanings' of nature and science. It complements the final chapter, eight, where I return to questions of meaning and politics – for our own time. Thus Chapters 1, 2 and 8 together provide a reflective frame for the core of this history. Some readers may prefer to focus first on the more 'scientific' material in the core chapters (3–7), returning later to consider the frame.

Please experiment, and use the book as suits you best. It is a set of tales, but also a set of tools; a pack of narratives that can be read in various ways, and a cluster of arguments about both past and present.

If you want more information, follow the bracketed references to the bibliography, where I have highlighted reference works and generally accessible texts. If you want to explore on-line sources, you could start with the links on www.man.ac.uk/chstm/

1

Ways of knowing: an introduction

L ET'S START WITH EGGS. Most animal species have
them, and so do many plants. If you so wished, you could
obtain details on thousands and thousands of different animal
eggs because naturalists have collected them, described them in
print and filed away the specimens and the data. In any of the
world's great natural history museums you could find out about
the eggs of the praying mantis, the chameleon or the elephant.
And for species with eggs of economic or medical importance –
for parasitic flies, edible fish or for hens – the data available is
overwhelming. How and why were all these specimens and data-
sets collected and kept? Who filled the filing cabinets of the
natural world?

And what of the eggs from which we as individuals began?
What do we know about them? In 1600, experts would have said
you came from a mixture of fluids – from the seminal fluid from
your father which acted on the seminal fluid of your mother.
Mammalian eggs were first described in the seventeenth century,
and so were spermatozoa – the millions of tiny 'animalcules' in
every drop of semen. What on earth, or in heaven, were they all
for? We are so used now to moving-pictures of sperms, eggs and
embryos, and of the late human foetus so 'lifelike' *in utero*, that
we may forget how all those microscopical or hidden processes
were made visible, how much ingenuity and interpretation those
'natural' images can hide. And that is just for the 'outsides' of
eggs.

Suppose you want to 'look inside'? Chemists can offer you
reams about the chemical composition of the eggs you eat; they

know them down to the level of the chemical elements. Biologists can tell you that the fertilised egg is a cell which divides to give rise to the millions of cells in the adult, and that the nucleus of each cell contains chromosomes which become visible during cell-division. Embryologists know about the 'germ layers' of vertebrate embryos and the ways in which a fertilised egg becomes a microscopical, hollow ball of cells; how one side of the ball folds in, like a flattened tennis ball; how the two-layered structure then grows so that the inner wall becomes your gut lining, the outer wall your nervous system and your skin, and a third, intermediate (germ-) layer produces the rest of your body. Most of this wonderful 'protoplasmic origami' was analysed in the nineteenth century under the leadership of German professors fascinated by 'development' in embryos, lives and societies.

By the end of the nineteenth century, they were trying to intervene. What happened if you removed half of a divided egg? Would you get two whole-embryos, or two half-embryos, or something in between? What would that experiment tell you about the 'potencies' of the parts of the egg? In laboratories by the sea, where the professors could spend their summers playing with marine eggs, they began to learn how to control the development of eggs. And now we have artificial fertilisation, cloning and 'Dolly' the sheep. In the twentieth century, embryos were further analysed to reveal that chromosomes are made up of nucleic acids, constituting genes that are switched on and off during development. Structural and chemical analyses now converge at the level of macro-molecules (Hopwood, forthcoming; Jacob, 1974; Needham, 1959).

This book is about all the various *ways of knowing* which went into 'eggs as we know them'. It is about *natural history* – the describing and classification of things; about *analysing* into various kinds of elements, like germ layers, cells and chemical elements; and about *experimenting* to control phenomena and to systematically create novelties. I will try to show that much of science, technology and medicine (STM) can usefully be understood in terms of these three ways of knowing and their interactions, and that this approach also illuminates the *histories* of

STM. My account includes planets and stars, clouds and cuckoos, minerals and chemicals, machines and steam engines, patients and bacteria, vacuum tubes and radioactivity, electronics and pharmaceuticals, atom bombs and genetic engineering – even languages and societies. It may not be comprehensive – indeed, it is conspicuously weak on mathematics – but it is wide ranging. In the next sections I introduce my method and say more about this particular approach to a huge topic – the history of science, technology and medicine in the West from the sixteenth-century Renaissance to the present. But first I want to add two further dimensions, which in some ways 'frame' the three ways of knowing – natural history, analysis and experimentalism – that I have just introduced.

I want to show how ways of knowing were linked with ways of *production* – to ways of *making* things, or to ways of tending and mending (in agriculture and medicine), or of defending or destroying (in military sciences and technologies). I am interested in the ways in which knowledge is built into commodities or other products, such as new pharmaceuticals or new weapons. I use the term *technoscience* for technological projects which are heavily dependent on science (or vice versa), and I suggest that it was chiefly from the later nineteenth century that (some) academics, (some) industrialists and (some) state agencies began to create the networks for the systematic production of novelty which were increasingly characteristic of the twentieth century. Eggs, for example, are now big business, not just through the standardisation and mechanisation of hen-farming, but in high-tech firms which clone the eggs of high-value cattle, using all the techniques of experimental biology and working closely with universities and with government agencies.

Technoscience, I will argue, is central to our world – to our artefacts and to our 'nature'. But we must read it aright. Why, for example, are the products of high-tech companies sold to us as 'brands', as symbols of (our) status, connoisseurship or modernity (Klein, 2000)? Their production may be technical beyond the understanding of most of humankind, but the appeal, manifestly, is a matter of 'human meanings'. We can only fully grasp these technologies if we understand the values; our appreciation of artefacts and of their invention and production must involve the study of *meanings*. But this is true for 'nature' also. *All* the worlds we know, whether through everyday life or 'science', have *meaning* for us. So they did also for the men and women who made our worlds and our understandings – but *their* understandings were not necessarily ours. To explore their *world-readings* and their purposes is part of the adventure of this book.

We can see that dimension easily enough by returning to eggs. When we watch a film of the developing embryo, it may suggest to us the possibility of modification and control – for better or worse. Few of us see there the glories of a divine creation unfolding before our eyes, though that attitude was the context and spur to much 'anatomising' from the Renaissance to the nineteenth century, and many of us still think it wise to cling to '*nature's* way', if only as 'precaution'. Abnormalities of human development are now the subject of 'dysmorphology', a branch of medicine recording and analysing 'monstrous births' in the hope of finding genetic or environmental causes. They are not seen as punishments or portents in the ways which once motivated such investigations – but stricken parents may still feel the power of that understanding when they ask 'why us?' And when popular genetics suggests to us that much of an individual's future is 'pre-formed', we might spare a thought for the seventeenth-century microscopists who saw tiny pre-formed beings hunched in the heads of sperms and thought they had solved one of the deepest quandaries of their (advanced) science. If God had made the world as a machine, how could one understand the 'generation' of new beings? Perhaps all beings had been 'laid down' at the Creation, stored one within another like Russian dolls. *That* was the meaning for them of the micro-mechanical men hidden in sperms (or maybe in eggs) and unfolding in the embryo. In a world of limited duration, perhaps all past and future generations had existed since creation.

And if you smile, spare a thought for the smiles of the future.

An outline of the method

My approach to the history of science, technology and medicine has four key features:

1 a long timescale: working through the history of the past 300 years;
2 breadth of scope: taking science, technology and medicine together – linking their history to other human histories;
3 dissecting science-technology-medicine into constituent elements – *ways of knowing*, with their different histories; and
4 presenting these *ways of knowing* as forms of *work* related to various ways of making and mending. In general, I use my 'ways of knowing' to bridge between esoteric, technical worlds and the worlds of 'everyday', both past and present.

Long timescale

Historians can be prissy about keeping to their periods. They worry that in studying events across centuries they may lose sensitivity to the peculiarities of particular times and places, and thus misread them. So we tend to specialise in particular periods and/or in the history of a discipline, e.g. the history of nineteenth-century physics. But when we work on a particular time and place, we can easily lose sight of the 'big picture', and the fact that we only know what something is when we know what it is *not*. To understand eighteenth-century medicine, say, and to communicate that understanding, we need to know how it differs from the medicine with which we live now. By knowing our contemporary medicine we can know more about the eighteenth century and vice versa. 'Periods' illuminate each other.

To be sure, we must be careful not to see eighteenth-century medicine in modern terms (the scholarly sin of 'presentism'), or to focus exclusively on the roots of later developments to the exclusion of aspects whose importance has since faded (the sin called 'Whiggism'), but in my view it is naive to pretend that we should or can forget our present categories. Could we do so, we would be no more use than anthropologists 'gone native' and unable to relate to fellow investigators.[1] To do justice to the past and to *use it in the present*, we need broad frames in which to think comparisons. For such reasons, this book takes a long period, and also a very broad definition of 'science'.

Wide scope

Most of us use the terms 'science', 'technology' and 'medicine' without much care and attention. Imprecision is endemic, confusions commonplace. When asked to name an achievement of science, we often name a device or process that might better be called technology. In turn, the word 'technology', rather like 'biology' or 'history', is ambiguous; it can refer to instruments and techniques or to the *study* of these instruments. 'Medicine' is easier to use (if we ignore the contrast between surgery and 'medicine' in the sense of 'internal medicine'). But 'medicine' in the wide sense sits oddly alongside 'science' and 'technology' since it includes the sciences and technologies that relate to health and disease. In that respect 'medicine', as an omnibus term, resembles 'agriculture', or maybe 'engineering' or 'electronics' – medicine can include anything from folk remedies to brain scanners, as 'electronics' spans from computer games to particle physics. These omnibus terms are very useful to historians partly because they avoid, or at least postpone, questions about what counts as science rather than technology. But, unfortunately, we do not have a full set of omnibus terms that can parallel 'medicine' – we still talk, for example, about chemical *sciences and technologies* – nor do we have a word that would encompass all these parallel science-technology fields. If we want an expression to include all of science-technology-medicine, in other ages as well as our own, I fear we must 'develop' one – which is why I lump science, technology and medicine as STM. This acronym does in fact have some present currency, both among librarians, and in the titles of academic departments such as mine – a Centre for the History of STM.

But that usage, and that convergence of historical studies is new. The histories of science, and of technology, and of medicine have usually been studied separately – often in separate faculties of universities, where historians have taught (for) scientists, engineers or doctors. Partly because of this separation, the conventional chronologies of the histories have been different, and our bringing them together forces a rethinking of their outlines and major features (Kragh, 1987). Whereas histories of *science* have tended to highlight the scientific revolution of the seventeenth century, most histories of *technology* have hinged (or begun) with the Industrial Revolution *c.* 1750–1850. Histories of *medicine* have been less 'revolutionary', though the 'birth of scientific medicine' is usually dated about 1870, and the 'Birth of the Clinic' in France c. 1800 has its supporters. This book is intended to help pull all these different histories together, to allow them to illuminate each other, and so to effect a synthesis that may work across most of STM.

Indeed, my synthesis extends beyond our current definitions of STM, for I follow where my ways of knowing lead. So under natural history I include archaeological remains and collections of pictures; my treatment of analysis includes economics and language studies, and a little on the social sciences, because these were continuous with analysis in the natural sciences and medicine; and in discussing the 'meanings' of the world and of STM, I explore some of the philosophical and cultural issues around STM, and the layers of human values that gave life to its projects. Science, technology and medicine are the focus of the book, but it has grown beyond them. Only by looking beyond STM can we see its limitations and its importance.

Dissecting out the elements of STM

My third key is dissection or analysis. As I tried to illustrate above, I take apart science, technology and medicine into 'elements' which I call 'ways of knowing'.[2] I focus on three ways of knowing, three ideal types of STM which I call *natural history*, *analysis* and *experimentalism*, but I also discuss a form of STM which I label *technoscience* and I explore the historical variety of 'natural philosophies' or *world-readings* – the *meanings* of nature and of STM.

The first three are ways of investigating nature or human artefacts. As noted, objects or systems can be described and classified (the activities I group as natural history); they can be taken apart by hands or just by brains, into their *elements* (this is the core of analysis); and the elements can be 'rearranged' to produce new and interesting phenomena (which will serve for now as my characterisation of experimentalism). I use the term *technoscience* primarily to refer to the production of 'scientific commodities' in academic-industrial-governmental complexes, such as the military-industrial complex or the medico-industrial complex. My term *world-readings* is meant to refer both to the 'decoding' of the world and to the systems of meaning found there. It covers the moral significance of eggs, the messages in the stars, the body as a divine creation or thermodynamics as foretelling the end of the world. It corresponds in some ways to 'natural philosophy', except that the world of human creations is included.

I will develop examples below, but we can note here that these ways of knowing are 'ideal types' in the sense of the German sociologist, Max Weber. To give a Weberian example – 'bureaucracy' was the rule-governed form of organisation paramount in government offices; the word could refer to a massive system of offices or to the workings of a few people, and offices could also contain other social types, e.g. friendship or charismatic leadership. Similarly, my ways of knowing can be used to characterise a multi-laboratory project (perhaps as chiefly analytical), or to describe its particular components. Many scientific projects will contain more than one kind of knowing; indeed, in my usage, analysis presumes natural history and experimentalism presumes both; technoscience projects typically include all my other ways of knowing. And by including a chapter on world-readings, I stress that all our knowings and all our creations are underpinned by systems of meaning and values.

We can go further and think of all these ways of knowing as potentially 'nested'. For example, experimenters commonly use analytical methods, and they also have to know the 'natural history' of materials on which they work. But note that nesting can occur in any sequence: relatively complex projects may have simple goals. We might regard the exploration of the moon as largely a matter of mapping or of natural history, but getting there involved lots of analysis and experimentation. Genomics – the spelling out of the human genetic code and the measurement of its variation – is best described as an analytical project, but its techniques involved much experimentation, and the collection of ranges of samples from the human gene-pool will draw on our natural-historical appreciation of the variety of humankind. Generally, STM projects are not to be 'dropped' into single ways of knowing, like specimens into boxes; rather, ways of knowing are to be used in analysing the *components* of projects and the relations between them. If you like, they are the *elements* of my analysis.

I will show that each of these ways of knowing has a history, and that these histories differ; new ways of knowing are created, but they rarely disappear. As Western society has grown more complex, so ways of knowing and doing have been built up. These 'ways' or projects interact in various ways and their 'coverages' vary over time. All my categories could be used, in principle, for any time and place, but as a matter of historical fact they became important at different times and in different combinations. In this view, history of STM is not a matter of successions, or the *replacements* of one kind of knowledge by another; rather it is a matter of *complex cumulation* and of simultaneous variety, contested over time, not least when new forms of knowledge partially *displace* old forms.

We will explore below the historical course of power relations between different ways of knowing, but an example here may help. I will argue in Chapters 4 and 5 that we might view the early nineteenth century as the 'age of analysis', by which I mean that analysis, both in my judgement and in the eyes of contemporaries, was the new, exciting, *dominant* way of knowing. But that does not mean that natural history was in decline, or that no one was doing experiments, or that nothing in that period

could usefully be described as technoscience. The contrary is the
case, as we shall see; in that period, as in all others, STM was
manifold.

With those preliminary comments, we can now survey the
ways of knowing (and later we can link them with ways of
making); at the end of this chapter we will see how these separate histories might fit into a chronological pattern. But if you
are growing tired of 'method', skim forward towards Chapter 2.

Natural history
'*Natural history*' like the classical Greek 'historia', compre-
hends the variety of things, whether human-made or natural,
'normal' or 'pathological'. I include the natural history of
animals, plants and minerals, but also phenomena such as the
weather, and artefacts – ancient and modern, exotic or industrial. In Chapters 2 and 3 I will discuss the creation of early
modern 'cultures of fact' and the social history of different
kinds of collecting, describing, naming and classifying; here I
want to stress two points.

First, the coexistence of the two dimensions suggested by the
very words 'natural history' – 'natural' suggests the taxonomic
aspect, the *range* of nature (or human creation), and 'history' the
biographical aspect of nature (or artefacts) *in time*. The taxonomic dimension is obvious enough – we shall discuss classification and displays from the Renaissance to the present day;
'biographical' in this context may be less evident, but it is important. Medical case histories are obvious examples of 'biography',
and throughout this book we will refer to biographical medicine
as a continuing tradition of seeing illnesses as disturbances of
individual lives; but towns could also be chronicled, and so could
'nature' – think here of Gilbert White's eighteenth-century
classic, *The Natural History of Selborne* – his collected letters
recording the changing seasons in the countryside around his
English village.

Secondly, I stress the continued importance of 'natural history'
and of 'information' more generally, even in the most technical
of sciences and of industries. Manufacturers and experimenters

need to know the range of materials available and the often subtle distinctions between them (Stansfield, 1990); they use many catalogues and much 'experience'. We live now in an 'age of information'; we feel excited, maybe overwhelmed, by the mass of 'facts' at our disposal. That perception may perhaps animate our appreciation of classifiers around 1700 who struggled with the huge variety of 'specimens' brought back by explorers and traders.

From the seventeenth century, at least, natural history has been the study of 'what we have' – in data banks, or public or private collections, or when we take our 'complaints' to the doctor and wonder 'what we've got?'. Pride of recognition, possession and presentation is often a motive force – whether for the 'cabinets of curiosities' collected by Renaissance princes, the huge national museums built by the imperial powers of the nineteenth century, or the moss specimens collected by knowing artisans of industrial Manchester. But, additionally, as we shall explore in Chapter 3, natural history in my extended sense was and remains crucial for trade and industry.

Analysis

If natural history records variety and change, *analysis* seeks order by dissection. I will argue that analysis comes into play when objects can be viewed as compounds of 'elements', or when processes can be viewed as the 'flow' of an 'element' through a system. Analyses can be of many types, even for the same objects of study. Thus, for some purposes, we can reduce minerals, such as rock salt or sea salt, to formulae such as $NaCl$ (sodium chloride); for other purposes we could 'reduce' them to their constituent 'unit crystals', or treat them as electrical materials and measure their conductivity. And processes can be represented by formulae, e.g. by saying that a steam engine achieves 50 per cent of its theoretical power output.

The divisions of natural history correspond to 'ranges' of natural things – plants, say, or birds – but each *analytical science* is constituted by the use of a particular kind of 'element'. Analytical chemistry reduces the whole world to chemical elements, thermodynamics to energy, and histology is the study of tissues in whatever animals they occur. These 'elements' were not obvious; in some sense they did not exist before 1750. They were 'discovered' and corresponding disciplines were 'invented' – mostly in the several decades around 1800.

In as much as analysis changes our understandings of objects, it may involve new kinds of classification and displays – 'deeper' than those of natural history. It is intimately associated with *comparative* work because common elements give new frameworks for comparison. In Chapters 4 and 5 we shall discuss the historical connections between analysis and collections, and the technical roles of museums, observatories, teaching hospitals and 'field stations'. I shall argue that the sciences of the early nineteenth century were primarily analytical, and that the new sciences were largely created by teachers of technical professionals (chiefly engineers and doctors), and that analysis remains a crucial, if much neglected, aspect of modern science (I also include sections on the human sciences that seem to me analytical, including political economy and the study of languages). All the analytical sciences were and remain hugely important for the improvement and regulation of technical processes. Most of the 'scientific practitioners' of the nineteenth century were analysts, variously working in industry, agriculture or medicine; that may be true for the twentieth century also. (Historians of science, unlike economic historians, rarely collect statistics; we scarcely have the categories.[3])

Experimentalism
If analysis is about taking things apart, *experimentalism* is about setting them up. The former is about specifying the composition of the 'known', the latter about putting together elements and controlling them to create new phenomena (or old phenomena in new ways).

My usage here is deliberately narrow. The early modern practice of collecting peculiar or impressive phenomena in 'experimental histories' I treat as part of natural history, part of cataloguing and displaying the world. 'Experimental measurement' is often best seen as quantitative analysis, whether in chemistry or physics, engineering or medicine; like other kinds of analysis, it is crucial to the refinement of technical processes. Experimentalism, in my terms, builds on analysis; it is about 'synthesis' and the systematic production of novelty. Without in any way denying the importance of earlier exemplars, I present experimentalism, in Chapter 6, as a set of practices chiefly institutionalised and theorised from the mid-nineteenth century, especially in universities newly oriented towards 'research' (rather than chiefly to the education of professionals).

Synthetic chemistry is a key example. Chemists who understood the elements (and structures) of a given class of chemicals could go beyond analysis. They guessed how to make complex compounds from simple ones; in some cases, the synthesised compounds were entirely novel. Physicists, creating a tradition of experimentalism, systematically explored the 'reactions' between such 'elements' as light and current electricity. We might argue that the creation of the phenomena of radioactivity, from the 1890s, was at once an extension of reaction chemistry and of this 'reaction' physics. But the example which I use most, and on which I based this understanding of experimentalism, was drawn from neither chemistry nor physics, but from the third great area of later nineteenth-century science – physiology. In his *Introduction to the Study of Experimental Medicine* of 1865, Claude Bernard (see Bernard, 1957) used his own researches on animal functions to show the potential of experiment in medicine. The text introduced, or at least defined, the 'control experiment' familiar to generations of scientists thereafter – when one repeats the experimental set-up in all particulars except the key variable, to try to ensure that the variable is indeed responsible for the effect in question.

And when the novel products of experimentalist laboratories could be developed as industrial commodities, then *technoscientific* complexes could be created and exploited.

Technoscience

Technoscience refers to ways of making knowledge that are also ways of making commodities, or such quasi-commodities as state-produced weapons. In as much as many governments over recent centuries have used and supported 'natural philosophers', doctors, engineers or indeed astrologers, technoscience is a 'timeless' category, a term with possible application in many societies and one that can be related in various ways to our other ways of knowing. Where state support and commercial goals were tied up with exploration and inventory, as in major national expeditions, it may be useful to think of a kind of technoscience in which *natural history* was the predominant way of knowing. Similarly, the interconnections around analysis were sometimes dense enough that we might usefully refer to *analytical* technoscience, for example in Revolutionary France or the industrial cities of early nineteenth-century Britain. But though state support was important to natural history and crucial to the formation of many analytical disciplines, I would argue that the interests of governments, of academics and of commercial companies were still substantially distinguishable from each other in the mid-nineteenth century, at least in peacetime. The technoscientific networks were still very marginal to government, to the direction of work in most universities, and to most industrial companies.

It was from *c.*1870, I will argue, that we begin to see the development of more inventive, intense and self-perpetuating synergies between the three sets of interests. Initially, of course, these technoscience networks were small; they involved natural history and analysis, but also, and crucially, they involved the systematic production of novelties by the interplay of experiment and invention. We may specify this form of technoscience as *synthetic*, and this is the formation I see as increasingly central to twentieth-century STM. The two initial networks remained crucial – one around the new electrical industries and the other around dyestuffs and pharmaceutical companies. States became involved as consumers of these scientific commodities and in some cases as producers, but also as regulators, especially for standardisation. Electrical systems, within and between nations, depended on agreed measurements and units; so did the use of new 'biological' treatments such as anti-toxins or vaccines. In as much as the products proved saleable or useful in war and public health, so companies, universities and states all had an interest both in developing further products and in reinforcing the 'innovation systems'.

The growth of technoscientific *companies*, producing new commodities from their own research laboratories *and* from the web of academic and governmental institutions, has been a major feature of the twentieth century. But my term 'technoscience' is also meant to cover high-tech projects funded and directed by state agencies, or even by state-backed academic consortia. Projects such as the US space telescope or the high-energy accelerator run by CERN (Conseil européen pour la recherche nucléaire – European Organisation for Nuclear Research) at Geneva, could be said to involve 'commodities' in the extended sense of products wanted by governments; they certainly involve dense intertwinings of universities, industry and government. For all such systems, trying to separate the science from the technology seems less profitable then recognising their characteristic, dynamic combinations. They all involve many kinds of natural history, analysis and experimentalism, and many kinds of making; they also involve massive organisation, and now dominate the academic and industrial worlds.

World-readings (or hermeneutics)

'Reading the world', by contrast, is as old as humankind. The various 'understandings' of nature and of STM will be explored in Chapter 2, where I sometimes refer to this decoding work as *hermeneutics* – a term that once meant the art of interpreting sacred texts, but which can be used more generally for the art of reading 'meanings', whether in texts, artefacts or nature. The products and frames of such decodings we might call 'natural philosophy' – if we extend the usual sense so as to include reflections on human creations and creativity, including the structure and processes of STM. For sure, these are not easy issues to pin down, especially at the start of a book. To discuss them, I borrow from accounts of science and religion and of science and literature. I also use work that is classed as philosophy and I draw on some 'political' histories of science and medicine, but there are no existing histories which cover the full range that I want to bring into focus.

I will begin with natural worlds that were meant to be read as systems of symbols, starting with the courtly worlds of the Renaissance. We glimpse how scholars interrogated texts, decoded the particulars of nature and used the clues to tweak the world by 'natural magic'. From this world of natural meanings I turn to northern, more Protestant worlds to see something of the (partial) 'disenchantment' of nature. The culture here is of navigators and merchants, of urban markets and coffee houses, and of aristocratic estates and churches in the country. It was a world created by God for men and women who also created, and who stood before Him as individuals, without the intermediaries of saints, cardinals or divine monarchs. We shall explore the cultures of accumulation and of agricultural improvement that were the context for natural history and 'experiment'. We shall also see how analysis was linked to navigation and surveying, and proved fundamental to the new 'cosmologies' of mechanisms and laws.

The 'world-readings' there established were mostly about *order* and *eternal truths*, but this book is mostly about *science and doing*. For this aspect of STM, the formative period is the decades around 1800 in France, Britain and Germany. There, we might say, were invented the three roles which made much of our modern world – the *technical-professional* (new kinds of engineers and doctors), the *industrialist* and the *researcher* (cf. Ben-David, 1971). Of course, such a claim needs massive qualification and explanation, so hold on to it for now and judge it later in the book. But the simplification may help us focus on the new 'meanings' of STM since the 1800s, and their interactions with older meanings. For example, the analytical sciences which I describe in the professional schools of post-Revolutionary Paris could be used for 'rule by experts', otherwise known as technocracy and still associated with France. Our attitudes to industry – the qualified welcome for the efficiency of rationalised production and for the creativity of inventors and industrial scientists, but also the worries over the mechanisation of people and the 'commercialisation' of everyday life – can be traced back to the Industrial Revolution in Britain. And the third role – the university professor – whether in physics or in the study of ancient languages, is rooted in the reform of German universities c.1800. Professors still inhabit our public worlds, though now we sometimes wonder by whom they are paid.

But just as the STM of the early modern period cannot be discussed without theology, so that of the modern period also requires attention to wider cultural frames – especially to 'readings' of human life in philosophy, literature and history, as well as in theology. The period around 1800 produced professors of philosophy and of literature, as well as of natural sciences. Even more important was the construction of a fourth new role – *the creative, individual, 'romantic' artist or writer* who taught new ways of finding meanings in nature and in human history. Romanticism will also be part of our discussion in Chapter 2. Like other ways of deriving meanings, and like other ways of knowing, the role of the artist can in principle be understood as a means of intervening in the world.

Ways of knowing as work

The fourth key to my history of STM is 'work'. I have tried to think of ways of knowing not just as mental operations but as modes of *work* – e.g., the work of analysing chemicals or pathological specimens, with all their characteristic locations, requirements, routines and products. In this respect my approach accords with much recent research in history of STM which you can find in specialist journals, but it is distanced from most of the general histories of science, even those written recently. These 'surveys' tend to focus on theories and ideas, such as 'evolution' or 'energy', and/or on the histories of particular disciplines, such as astronomy or chemistry, rather than on the forms of practice which may extend across (and help constitute) many disciplines.

By concentrating on the work of science, I hope to bring history of STM closer to social and economic history, and to facilitate studies of the kinds of scientific work that were *concurrent* at any given time. For example, we could in principle ask how many 'analysts' were at work in the universities and industries of late nineteenth-century London or Boston, or how the labour forces of natural history grew over the century. Such histories would be quantitative, like some studies of industrial history; more generally, they would be about synchronicity and interdependencies – not just about novelties and new frontiers (Edgerton, 1999). Though, for the most part, such histories lie in the future, thinking in these terms is useful now, not least to deepen our understandings of change.

Scientific change involves *changes of work patterns*, not just of ideas, so the *institutions* of STM are a key part of this book. For example, the emergence of 'clinical examinations and pathological anatomy' in Paris after the French Revolution, was much more than a theoretical shift in medicine; it involved major changes in professional and educational structures, in the control of hospitals and in the work routines of doctors. It could not have been realised without the political changes and, had those changes been different, then *different* medical aspirations would have been put into practice. Similarly, the rise of experimentalism in physics as a regular practice and as a social institution had many and various preconditions, including advanced and systematic education, analytical concepts and practices, workshops and laboratories for the construction and use of experimental systems, plus the time and money needed for would-be experimentalists to devote themselves to these creations, rather than to the demands of, say, commercial analysis.

It was in exploring such historical relations between knowledge and practice, or science and technology, that I came to think about three ideal types of *making* which I call *craft, rationalised production* and *systematic invention*. At first I thought of them as a way of explaining my ways of knowing, through the analogy of knowledge production with manufacturing. For example, we can use the three 'types' of making to reconstruct the history of manufacturing in terms of displacements rather than replacements: the Industrial Revolution around 1800 could be seen as based on rationalisation and mechanisation of production, while recognising that crafts continued to be important (Braverman, 1974); systematic invention, I suggest, became important at the end of the nineteenth century, when industrial research laboratories were created for that purpose. My ways of *making* functioned in similar ways as my ways of *knowing*, and they also 'nested' similarly.

It still seems to me that the ideal types of making may be a useful tool for the history of technology – not just for the history of manufacturing, but also for agriculture, medicine and the military. Craft, rationalisation and invention can be ways not just of making, but of tending, mending, defending or, indeed, of killing – think of keeping a pet hen, of battery-farming chickens, and of genetically engineering new breeds of poultry; or of duelling, trench warfare and guided missiles (and of all their possible coexistences and interactions!). But I came to realise that the relationship between these ways of knowing and ways of making was much more than analogical.

I would now argue that there are close, systematic linkages, recognition of which helps illuminate many common questions about science–technology relations. It is not hard to see that rationalised production, which typically involves the deconstruction of craft activities and their reconstruction as machines, depends on an 'analysis' of craft work similar to that involved in the deconstruction of chemicals into their elements. And I have probably already said enough about my perspective on 'experimentalism' for you to see the close kinship with 'invention': both create novelties from elements – experiments are usually for enlightenment, inventions usually for profit. Indeed, experiment and invention, like analysis and rationalisation, might be viewed as two sides of a single coin. But what of craft and natural history? – not a combination much discussed by historians.

It would be trite to claim that all natural history involves craft, though it is *not* trite to stress that all ways of knowing involve crafts (Polanyi, 1958) and several recent sociologists of science have stressed the role of tacit knowledge and skills, even in mathematics (e.g., Collins, 1992). It is more specifically enlightening

to note that most early modern (and many late modern) technologies, e.g. agriculture, textiles and medicine, involved much
special knowledge of minerals, plants and animals, as well as
craft skills. Although these characteristic alliances have been
hidden by historians who concentrated on industrial technologies, they are coming to light in newer accounts of the 'scientific
revolution'. We note, for example, that most museums contained
both the works of God and the works of humans. But we can go
further. Craft products, I will argue, were commonly seen as
quasi-species in much the same way as animal and plant species.
Think of the *local* varieties of cloth – the 'blue denim' of 'jeans'
was once 'Bleu de Nîmes', a variety of cloth made near Nîmes in
the South of France, with characteristics peculiar to the place; or
of wines such as 'Bordeaux' or 'Champagne', and the ways in
which they depended on (and helped constitute) both region and
season. Vernacular housing or traditional kinds of spades show
much the same pattern, and like cloths and wines they could
suffer from characteristic defects and 'diseases', as we shall
explore in Chapters 3 and 4.

Thus, summarising very tightly the relation of knowing and
making, we can see both 'extended natural history' and crafts as
aspects of material culture – including the creations both of
people and of nature; they concern the 'thinginess' of life – gath-
ering and making, possessing and displaying. Analysis and rationalisation are employed both to 'deepen' classifications and
diagnoses, and to refine and regulate technical processes, especially in medicine, agriculture and industry. Experiment and
invention are aspects of the search for novelty and the creation
of model worlds; they can be commercialised or put to state uses
by technoscientific networks that manufacture 'knowledge commodities'.

My setting out these claims about ways of knowing and ways
of making is itself a novel enterprise. It may seem strange to
readers at this point; indeed I feel myself that the historical
project is only just begun. But I hope thereby to open up new
avenues by which our histories of science, technology and of
medicine can be combined in new and fertile ways.

Missions for this book

If we take together my four keys – breadth, history, dissection and work – we begin to see the project of this book – to present the STM as a variety of changing projects within the economic, political and cultural history of the last few centuries. By analysing STM we can show how its constituents are part of those wider histories and thus part of our varied heritages.

Ideally, I would include other traditions than the Western, but this I must leave to others. I concentrate on Europe and North America over the last 300 years, and especially on Britain and France, but even within that frame the coverage may be thought very uneven and excessively Anglocentric. I am aware that like so much history of STM, my account says relatively little about military STM. I have missed out much that I could have included, for this is an introductory text not a comprehensive survey, yet, even so, I have ventured into many areas outside my special fields of study, and doubtless I have made mistakes. The text is exploratory and provisional – introducing a method and trying it out by raiding, rearranging and sometimes revising the works of my fellow historians. I shall be pleased to receive corrections and suggestions.

In many cases I have used local, even personal, histories. I have often drawn my examples from Paris and Manchester, the places I know best. The importance of those two cities in the Age of Revolutions (French and Industrial) will be evident; so, I trust, will the virtues of comparative history; but I have a further reason for concentrating on them. By selecting different examples from the same places, for different times and topics, I hope to give the reader some feel for the 'local connectedness' of developments in STM. Indeed, I would like to think that readers will be encouraged to think about that connectedness in the cities and countries about which they know and care. I love local history; I have written a very general book, in part to facilitate such studies.

I sometimes use personal examples for similar reasons – to encourage historical reflection on how STM 'plays out' for each of us now, and to show how the work of academic historians of science can itself be understood historically. The final chapter is unashamedly local, in time and space. Even more than the rest of the book it draws on British material, and it clearly arises from the public politics of STM around the year 2000. I am told it will date quickly and I tend to agree. But even when dated, it may still serve as an example of how history may be used, and why it is created. I very much wish that the later twentieth-century sections of this study formed a better bridge between historical re-creation and reflection; but even so, it seems worth making the links explicit in the hope of increasing the utility (if not the consistency) of the whole book.

Maybe it is also appropriate to spell out here the audiences for whom I hope the book will be useful. I envisage three overlapping groups:

1 historians and other analysts of social and intellectual change;
2 scientists, engineers and doctors, and all the other women and men who are professionally concerned with the increase and use of expert knowledges and techniques; and
3 the general reader – or perhaps the general reader in all of us – the citizens and consumers, patients and politicians, merchants of change and champions of conservation who seek better understandings of the roles of expertise in our world.

It is not easy to address the needs of all these groups simultaneously, so please feel free to skip those of the next three sections which concern you least. If you have never heard of the historians I am about to discuss and/or you are not professionally concerned with STM, do not worry – the book does not require such knowledge, so fast forward now to 'For the general public', and ignore all the bracketed references in the text.

For historians

For historians, especially for researchers and students in the history of STM, I hope this book will serve as a tool kit and a new map. It draws on recent studies which are valued in these communities of scholarship, and no doubt it will be judged as to how it 'fits' with them. I like to think it will enrich the best of them, enhancing them by comparisons and contrasts – by an oblique light. Like all works of (analysis and) synthesis, it depends on the work of hundreds of scholars. History of STM, in its various guises, has thrived in universities since the 1960s, especially in the USA and Britain; this book is possible only because of that innovative (and collegial) scholarship, and some of my debts are recorded in the Bibliography. But this is not the place for the detailed analysis of recent work,[4] rather I now introduce some of the authors now dead or long retired who provided intellectual sustenance and inspiration: Michel Foucault, Thomas Kuhn, Lewis Mumford, R. G. Collingwood, Owsei Temkin and Max Weber – major figures all, or representative, but in very different ways and rarely for the same audiences. The texts I have found most useful are listed in the Bibliography;[5] here I give a few notes on how I have used these authors.

Foucault is a mainstay for medical and cultural historians, Kuhn for historians and sociologists of physical sciences. I use Foucault chiefly for his analysis of 'the archaeology of knowledge', for his explorations of the different 'epistemes' (or ways of knowing) which he presented as fundamental to each period of the modern world. Kuhn is best known for his idea of 'paradigms' – models of process and outcome which can be followed by scientists in any particular discipline so as to gain (and recognise) correct results;[6] pursuit of paradigms constituted 'normal' science; scientific 'revolutions' were prompted by the breakdown of paradigms. These ideas are helpful, especially for the period around 1800, but in this book I draw less on Kuhn's *The Structure of Scientific Revolutions* (1962) than on an article where he tried to map out the history of STM, from physics to biology, from the seventeenth century to the nineteenth (in Kuhn, 1977). His distinction there between empirical and mathematical traditions is one of the roots of my distinction between natural history and analysis.

But neither Foucault nor Kuhn has much to say about technology; here my favourite 'older' author is Lewis Mumford, a disciple of the biologist and town planner, Patrick Geddes, and a representative of a critical history of technology first developed between the World Wars, drawing inspiration from that most astute historian of nineteenth-century technology (and Manchester) – Karl Marx. From the inter-war period too, comes the modern tradition of history of medicine, still represented by Owsei Temkin – developed in Germany, transplanted to the USA in the 1930s, submerged somewhat in the positivistic 1950s, but re-emergent from the 1960s as a new generation sought to assess the relations between medicine, culture and society. It was an émigré contemporary of Temkin, Erwin Ackerknecht, who introduced four 'types' of medicine seen in history – book medicine, bedside medicine, hospital medicine and laboratory medicine (Ackerknecht, 1967), which were later developed and popularised by Jewson (1974; 1976). These served as my starting point for the historical analysis of science and technology more generally.

My debts to Collingwood and Weber are more distant and diffuse. To Collingwood, for the tradition of history as understanding which he represented almost single-handedly in Britain between the wars; for his discussions of magic, craft and art which seem to me much more useful than their present obscurity would suggest; and for his childhood reading of antique science books by Lake Coniston, from which he learned that science was made like other human activities – over time. To Weber, albeit indirectly, I owe the chief method of this book – the use of 'ideal types'. I use ideal-typical 'ways of knowing' in the way that Weber used 'bureaucracy' as a type of authority – and indeed as a way of knowing and a way of working. It may seem odd to make the acknowledgement here: the method is widespread, it did not originate with Weber and I have not studied his writings in detail. But he stands as a model for his range and his clarity; and we need those qualities as targets, however distant (Weber, 1949).

By using Weber's method I have tried to create a typology of knowing (and making) which can be used across the whole of STM to explore concurrent patterns of work. I will argue that we can characterise periods by the relative importance of different ways of knowing, and I will attempt explanations of changes *between* periods. But I also want to stress the 'longue durée' histories of various ways of knowing; revolutionary changes in science may *displace* previous ways of knowing, but they do not wholly *replace* them. Between periods, places and disciplines, my method is analytical and comparative – finding elements in common, showing how they developed and how they were differently arranged and structured. At the end of this introductory chapter I will sketch the outline history of STM which seems to me to arise from this method of analysis, showing how the usual 'periods' may be characterised, and how we may trace patterns of succession in 'newly dominant' ways of knowing. But it is central to this book that science, technology and medicine *are and have always been* much more *plural* than most people know; at any time there have been lots of different ways of knowing and of making.

This plurality also applies to *historical* studies, in the past and now. There are lots of ways of doing history, because it serves many purposes and many audiences. (There may be lots of ways of using my typology to produce outline histories different from that presented here.) Yet professional historical studies, like the various components of STM, have their own rules – not 'anything goes', not all comparisons make sense, not all views of the past will stand the tests of evidence which historians use, any more than all theories or inventions withstood the tests of analysis, experiment or use. Critical pluralism (rather than relativism) is central to my account of STM and to this practice of its history. I hope this book will serve many kinds of historian in critically developing their own approaches.

For practitioners of STM

For practitioners of STM I hope to provide an aid to reflection and action – a means of situating their own concerns and problems within a large historical and contemporary frame, which can also include such other 'cases' as they know from experience, anecdote, reportage or historical writings.

Suppose, for example, we ask what a biologist might draw from the excellent biographical studies now available on Charles Darwin, or of Lamarck and his evolutionary theories *c*.1800 (Browne, 1995; Corsi, 1988; Desmond and Moore, 1992; Jordanova, 1984). Good historians no longer see such figures as simply 'contributors' to a supposedly continuous history of speculation and theory about animal evolution. They stress context, and the way that scientific problems were 'of their time'. But how many biologists, or general readers, could use such accounts in constructing a history of science in which they themselves also had a place? Scientists are still taught to think of science as accumulated truths, and to measure past scientists thereby. Though the biographies elucidate the particular connections between science and 'period', such links will remain anomalous for much of the readership. Part of my aim here is to make *many* such links, to give general characterisations of periods, linking social and political histories with the then 'dominant' and 'recessive' ways of knowing, so that particular developments can be 'placed', both as types and in a general historical overview. By using a simple typology and advancing a new, diagrammatic model of the history of STM, I hope to create accessible frameworks in which 'cases', lives, institutions and disciplines – in all their richness – can be more easily related to each other and to wider histories. The method is simple but not, I hope, constraining. Any single 'case' can be elaborated indefinitely. I do not wish to reduce the complexity of history, but to offer maps and methods of mapping that can be used at various scales – from bird's eye views to close-ups.

These frameworks should also be useful for that part of history which is not yet the past – for the analysis of our own times. How are different *kinds* of scientific work related to each other and to other kinds of work in our present? How might we think about the role of analysts or of 'information gatherers'? What kinds of intellectual and practical infrastructure are required by various sciences and technologies? Are new ways of knowing emerging with new technologies? Does it make sense to talk about 'the scientific method'? What is gained or lost when researchers refer to themselves as 'scientists'? (Could it be that now the effect is rather as if office workers took to referring to themselves as bureaucrats and organising national and international associations for the advancement of bureaucracy?) I particularly hope that the book will appeal to doctors, engineers and others who are sometimes described as 'applying' science.

In this text the notion of 'applied science' will appear only as an actor's category – a term used by the people we are studying. Though I acknowledge that many people have used the phrase and acted on its message, I think there are better ways of analysing these concerns. I suspect that in Britain and maybe elsewhere, clinical and engineering investigations have been handicapped over the last century by having to operate with a model of science that did not fit *their* customary kinds of research. Doctors and engineers often focus on analytical studies rather than 'laboratory' experiments, but if funding committees equate science with laboratory science, and especially with experimentation in a narrow sense, then research 'nearer to practice' may be marginalised. We may hope that a wider view of 'science' may help promote a more varied research culture.

For the general public

Since the 1980s British scientific establishments have been much concerned with the 'public understanding of science'; most of the consequent activity could reasonably be described as public relations (PR) on behalf of scientists. But it has become clear, even to the PR agents for science, that the public is not to be thought of as vessels more or less full of science; it is more profitable and respectful to see our fellow citizens as agents pursuing their own projects, with various resources and constraints. Here is a theme I want to develop in this book. Indeed, I would count it a success if it increases people's ability to relate esoteric knowledges and technical products to the lives they wish to lead, individually and collectively.

I hope to enhance the public utility of history of STM as a means to these private and public ends – to support the roles of citizens whether as 'consumers' of STM or as its 'backers', giving to research charities or paying tax to support universities and other state-funded research. Of course, these two sets of roles are linked, but they are worth distinguishing for analytical purposes, not least because of the growing tendency to see citizens simply as consumers (not just of market goods, but of state services and even of political policies). Remember here the sadly surprising response of a national art gallery director when asked about entrance charges: 'the public come in free because they own these paintings'.

In as much as we are all 'supporters' of STM, we need to focus on the choices available to us as individuals, and via collective political machinery. To do so, we need estimates of the plurality of STM, of the possibilities of emphasising some aspects rather than others and of the potential efficacy of public policies related to STM. Generally, we need to fit the meanings of STM into our individual and collective views of politics.

History can help that enterprise by providing both tools and frameworks. I have already introduced my tools – my ways of knowing and my ways of making; I end this rather long introduction by sketching the framework. Each of my main chapters is devoted to the history of a particular way of knowing, such as natural history or experimentalism; the book is 'chronological' only *within* the chapters, so I here outline the overall history that emerges from my use of this mode of analysis, extending the sketches I gave earlier in this chapter. This quick survey is meant as a preliminary orientation. It may seem rather too dense for digestion, especially if you are new to the subject, but it may prove useful for reference as you make your way through the book.

An outline of the story

The coverage starts, at least in the early chapters, with the Renaissance of the sixteenth century. I would argue that its astronomy can best be seen as an analytical science, inherited from the Greeks and pursued by mathematicians; its natural history was part of the Aristotelian tradition of the universities, but usually subordinate to a natural philosophy which saw the world as a purposive organism. More generally, and especially in the tradition named Neoplatonic, the world was read as if a text – it was full of hidden meanings – a drama; for example, monstrous births were portents.

Histories of science written after the Second World War usually present the 'scientific revolution of the seventeenth century' as *the origin* of modern science – as the birth of our mechanical cosmology and of 'the scientific method'. This book is much more pluralist – it is about many contested developments at many times, and about many methods. In my account, following Kuhn,[7] the seventeenth century saw two complex and related shifts – the (partial) 'disenchantment' of nature and the (partial) restructuring of the mathematical sciences.

Especially in the Protestant cosmologies of the seventeenth century, God and humankind were alike *distanced* from nature. God had created the world and was not bound by it; humans studied God's creation – for its fullness and its regularities, no longer for particular messages. Natural-historical studies acquired a new status independent of philosophy; and human craft-creations could serve as models of the nature created by God; and nature, like man's creations, could usefully be interrogated by experimental manipulation. My account of the seventeenth century, like recent studies of Holland and Britain, features the workaday worlds of medicine and crafts, catalogues and recipes, exploration and collections (Jardine, 1999).

Such accounts tend to marginalise what was once seen as most 'revolutionary' – the developments from Copernicus to Newton. These I sketch primarily as revisions of classical analytical sciences, notably astronomy – which were then combined with the new analytical mechanics into the great synthesis of the *Principia*. These developments were fundamentally mathematical, beyond the comprehension of most educated men and women, even of the savants or medical graduates who promoted natural history and looked for new 'experimental' phenomena.

They were, however, associated with more general and accessible shifts in the natural philosophy of such savants: crudely, the Greek world-as-organism was partially displaced by the world-as-machine. That new model fed the appetite for analysis; savants tried to understand chemical reactions, animals and societies in mechanical terms, but with limited success.

The knowledges of the eighteenth century can thus be seen as in part analytical – the 'exact sciences' that were more or less reduced to mathematics and mechanics, plus the more or less speculative natural philosophies which imagined how the rest of nature *might* be 'mechanised'. But for the most part, accounts of nature were in fact natural history – both taxonomic and 'historical'. Naturalists and physicians, savants and antiquarians described their surroundings, possessions and ailments, and followed them over time in local chronicles. The eighteenth century was the great age of classification – not just as a tool, but as a model for all knowledge. Indeed it was partly in seeking deeper means of classifying that the new analytical sciences were created from the end of that century.

I shall argue that these new achievements of analysis owed much to the entry of professional teachers into state museums, teaching hospitals and engineering schools, especially in France. In such professional contexts, in the industrial cities of Britain and in the reformed, 'research'-oriented universities of Germany, new analytical disciplines were created. They were crucial for the refinement of technical practices; that was major reason for their invention and for their flourishing in an increasingly industrial world.

I then explore the creation of systematic experimentation around the mid-nineteenth century – on the back of analysis, but not initially in the professional training institutions or museums. Experiment was about making and displaying new worlds, it was somewhat distanced for professional practice. It was institutionalised in more theatrical sites such as the Royal Institution in London, and it was systematically built up in universities with an ideology of 'research', especially in Germany. I will argue that such experimentalism was often presented as fitting for industry, as education for 'changing the world'. From about the same time, its more worldly twin – systematic invention – was pioneered in the USA and in Germany. Together they came to be exploited as (industrial, synthetic) technoscience.

In this book, the twentieth century appears chiefly in terms of developments of nineteenth-century patterns, especially the spread of technoscientific interactions. I discuss the industrial-academic-governmental complexes which produced electrotechnics, electronics and informatics, and the growth of the pharmaceutical industries. I outline some of the wartime developments, and I survey the growth of state-sponsored STM after the Second World War, including the military-industrial and medical-industrial complexes. Finally, I note the late century shifts in the driving forces of technosciences – away from state power and towards commerce.

In the final chapter I attempt to use my historical analysis to reflect on the dynamics and problems of STM around the year 2000. Drawing heavily on British material, I explore the commercialisation of science and related issues of 'public understanding'. I try to use my typology and history to illuminate the current politics of STM, and I ask about the *meanings* of STM, of our 'nature' and of our creations. Thus at the end of the book we shall return to the issues we take up now in Chapter 2 – on *readings of the world*.

Notes

1 For the contrary view see Ashplant and Wilson (1988) and Wilson and Ashplant (1988); for an expansion of my argument see Pickstone (1995).

2 The method was first presented in Pickstone (1993a) and elaborated in various ways in Pickstone (1993b; 1994a; 1994b; 1996; 1997).

3 But see Price (1965).

4 For a recent introduction and further references see Golinski (1998).

5 As guides to these authors (on some of whom the literatures are

enormous) see, for Foucault, the biography by Eribon (1991) and the excellent primer on his early works by Gutting (1989); on Kuhn, Barnes (1974; 1982); on Mumford, Miller (1992); on Collingwood, his autobiography (Collingwood, 1939) and Krausz (1972); on Temkin, the Introduction to his selected papers (Temkin, 1977); and on Weber, the biography by Bendix (1960), Hughes (1974) and many collections and commentaries.

6 As we shall note below, the analytical sciences I describe from the period around 1800 provide most of the good examples of Kuhnian paradigms – e.g. Lavoisier's chemistry. I owe to Andrew Warwick the suggestion that the conditions for the creation of these analytical disciplines, e.g. in professional schools, were the conditions Kuhn evoked, rather unhistorically, for paradigms generally. And indeed, that the period around 1800 was the first in which such conditions could be found. In this view, only thereafter could any science be 'normal' in Kuhn's sense.

7 Here and elsewhere my debt is less to Kuhn's widely known work on paradigms and scientific revolutions (Kuhn, 1962) than to his exploratory essay on 'Mathematical versus experimental traditions' in his *The Essential Tension* (Kuhn, 1977).

2

World-readings: the meanings of nature and of science

THIS CHAPTER IS PRIMARILY a historical review of the ways in which 'nature' and 'science' have been understood in the West, but I begin by underlining the *range* of such understandings in the present. Two anecdotes about understanding 'ourselves' may serve to introduce these themes.

A German historian has published a wonderful book on eighteenth-century medicine, based on case notes left by a country doctor. She shows how flows of fluids – of blood, milk, sweat and urine – permeated the imagery of the doctor and of his patients. How different, she remarks, from our own times when patients go to doctors to discuss the possibility of a myocardial infarction (Duden, 1991). Well, maybe in Manhattan! But where I come from they still have pains and tensions, weakness and 'the runs', even if they know some pathology.[1]

A former colleague is an expert in Chinese medicine and its history. Since university rules required that the students' essays and examinations be 'second marked', and since we had no other expert on Chinese medicine, I was the second marker. The students' accounts of Chinese medicine were predictably rich, but I was bothered by their accounts of Western medicine that served, implicitly or explicitly, as foil or point of contrast. The Chinese were allowed to worry about balance and imbalance; *we* were supposed to think in the pathologies of modern medical science, and as good Cartesians we were supposed to separate mind from body. Sometimes, I wondered whether the description of the Chinese approach was more accurate for ordinary Western patients, even *c.*2000, than was the account based on 'scientific medicine'.[2]

These instances are not unusual. When historians and/or anthropologists examine 'other cultures' they inevitably rely on an explicit or implicit understanding of the 'home' culture shared by author and readers – and these accounts 'of us' are often simplistic in the extreme, far cruder and less pluralist than the accounts of 'the other'. This seems to me a serious problem. Partly it arises from the feeling that we know our own culture just by living in it – and so, in a sense, we do; but if we are not careful, the richness of that tacit understanding can be lost when comparisons are made explicit. Then we may adopt Western 'self-images' devised to *contrast* with other cultures (other peoples, other times), rather than to facilitate sympathetic comparison. We may equate the self-understandings of Western patients, say, with the workplace methods of (some) professional health workers, forgetting the huge *variety* of understandings which can be found in any modern hospital. One purpose of this book is to facilitate appreciation of that variety. That is why I have introduced the method of ideal types so as to point up the different ways of knowing, their histories and their relationships. We can carry forward our exploration of this method by surveying and dissecting our own, everyday, shared knowledges. Medicine, here as elsewhere, serves as a good model for other aspects of STM.

Variety in modern Western medicine

At this point, you could try an exercise: jot down a few ways in which you might think of a sore throat. Then enlarge the frame to include the ways in which doctors, or scientists or pharmaceutical companies might think of it.

It could be the result of talking too much, or the consequence of going out last night to a smoky café or walking home in the cold. It could be a family weakness inherited from your mother. Maybe you always get a sore throat at this time of year. No doubt it will clear up in a few days if you take care, keep warm and take lemon and glycerine or Scotch whisky. But suppose it does not clear up as expected. You may become worried about more serious conditions that run in your family, or about your failure to attend to warning signs. You may feel that the prolonged affliction is a judgement on your behaviour, or a reflection of some deeper problems in your life. Maybe you need to get yourself sorted out, maybe a therapist could help. It is increasingly obvious, even in Western societies, that not all health-related expertise is to be classed as medical.

Or you could go to your doctor? He or she can tell you what infections are 'going around' and suggest the type of infection you may have, and its likely outcome. But maybe the doctor will also take a swab test and send the sample to the laboratory in the local hospital; it will report whether or not your sore throat is due to streptococcal bacteria. If it is, or maybe in any case, you could be prescribed antibiotics, to be collected from the local pharmacist. (And if the infection were a rare or serious one, the doctor would send data to the Public Health Department.) The pharmacist may be the same person who sold you the lemon and glycerine. Indeed, the shop displays a whole range of commercial remedies for sore throats, but behind the screen there is another range of 'ethical' preparations which can only be obtained on prescription. You cannot help thinking how sore throats are a boon to pharmacists and pharmaceutical manufacturers, but you also get to wondering about the difference to your prospects which you owe to the discovery of antibiotics. Where did they come from?

Outside town there is a large laboratory belonging to a pharmaceutical company. They still do research on new antibiotics, not least because bacteria are becoming resistant to the older forms. They synthesise new chemicals and try them out on bacteria on plates of jelly, or by using laboratory animals – the 'beasts of burden' that carry infections for humans and spare them from the initial trials. If a new antibiotic works well, clinical trials will be run, often in a teaching hospital staffed by expert physicians, working closely with university departments of medical science. The university scientists are looking for discoveries that will lead to publications and academic reputation; the doctors may be credited with a new cure. The pharmaceutical managers hope to make a profit, and to that end will employ a sales force to market the drug in the medical press and by visits and 'promotions' in hospitals and general practioners' surgeries.

In our complex society, your sore throat has many meanings: a judgement or a sign of misdoing, an instance of a disease species or a kind of inflammation, a possible public hazard, a product of a bacterium, a process that has been imitated and studied in laboratory animals, and an opportunity for the makers of medicines corresponding to one or more of these understandings. Suppose we ask about the history of this complex of understandings. Usually we will be told the 'scientific' history, roughly as follows.

Louis Pasteur, a French chemist, argued in the 1860s that fermentation and putrefaction (thought to be akin to 'fevers') were caused by simple microscopic plants – let's call them microbes. By the 1880s, these microbes could be 'cultured' on nutrient jellies, and so identified and reliably reproduced. By culturing microbes from animals and introducing them into other animals, it was shown that several key infectious diseases were 'caused by' specific kinds of microbes. Microbes could be attacked by antiseptics, if they were 'external to the body', but antiseptics damaged the body's own tissues, and so medical scientists searched for 'magic bullets' which could kill bacteria inside the body. Pharmaceutical companies invested heavily, and by the 1930s agents were known which would cure septicaemia, etc.

It was also known that some fungi could inhibit bacteria in cultures. During the Second World War a team of researchers in Oxford took up one of these fungi and showed that a stable agent could be extracted (penicillin) which was remarkably effective when injected into humans suffering from common bacterial infections. Pharmaceutical companies (first in the USA) worked out how to produce penicillin in bulk, and also invested in searches for other fungi with similar properties. By the late 1950s, several such agents were in common use, and all common bacterial conditions (including tuberculosis) could be cured. But by the end of the century it had become clear that bacteria were not so easily vanquished; new strains had emerged which were resistant to most antibiotics. Diseases came to be seen as a matter of 'balance' between the adaptive powers of microbes and of humans.

But why give the history only for 'scientific medicine'; don't the other views have histories, or at least, distributions? What can we say about the histories of meaning, or about the histories of 'sore throats' as an official category or as 'category of nature', and one that might vary between countries (as the significance of the liver varies between European countries)? Most cultures know how to read diseases for 'meanings' and many cultures have ways of treating illnesses as 'matters of fact', akin to plants and animals in having their own characteristic forms and durations. It is tempting to assume that these forms of understanding are less important to us because they are more widely shared, or that they are less bound up with Western history because they are also part of other histories. But that may be the reverse of the truth.

Much of this book will be about the history of these simpler understandings: about *natural history* in a very general sense, and about meanings. Natural history – the arts of describing and classifying which underlie all our more complex forms of 'naturalistic' knowing – will be the subject of the next chapter. The rest of this chapter is primarily about '*readings of the world*' or hermeneutics (Ferraris, 1996), about the meanings which have attached to 'disease' and 'nature' and also to the processes of 'science', in other times and places.

Meanings and readings

Most cultures exhibit symbolic understandings. It is a commonplace of medical sociology and medical anthropology that our culture is no exception. People ask 'why me?' 'Why this sore throat?' These symbolic understandings involve 'readings' of diseases as *message bearing*, they involve decoding as if diseases were texts, and they involve the supposition that the diseases are part of a 'meaningful' cosmos (Cornwell, 1984; Helman, 1986). There is no originality in my pointing to these aspects of our culture, but originality is not the only end of scholarship, and some commonplaces bear repetition, even emphasis, because they are often 'missed out' of standard accounts. Especially in the history of science and technology we need to remember that understanding 'things' as things is *not* the basic currency of our individual and collective lives. Such 'naturalistic' understandings are very important aspects of culture, and they have been remarkably elaborated in Western cultures, especially over the nineteenth and twentieth centuries; but, at root, we live in a world of 'agents', meanings and feelings, and of communications (face to face, distant, or imaginary).

In this world of purposeful actions and meanings we may include non-humans, sometimes in ways we now class as allegorical or metaphorical rather than 'real'; we may talk to trees or learn to trust a climbing rope. If asked to explain such 'communing', we now resort to pop psychology. We would excuse our talking to trees as a means of focusing our own spirits rather than theirs. Or, rather, some of us would, for some purposes – when we are operating in modern-Western-knowledge mode. But where did that mode come from? Whence the divide between *our* minded bodies and the trees' mere wood? Why is the 'meaning' of my sore throat a matter for my friends or therapist, rather than my doctor? Those divisions have histories.

As we see so clearly now in debates about birth, death or animal rights (and eggs), individuals and cultures vary in the boundaries they sketch between 'agents' and 'mere objects'. But whenever we *contrast* Western 'naturalism' with non-Western 'symbolic understandings', we get off on the wrong foot and introduce a basic asymmetry that is unhelpful, both intellectually and ethically. For even when we deal with the most 'modern' aspects of Western culture, 'readings' and hermeneutics must be to the fore. How else shall we understand modern commodity cultures and their advertising? It is no accident that public relations firms employ anthropologists; and if magic is the technology of meanings, or the art of raising or depressing morale (as

Collingwood, 1938, maintained), then are not advertisers, PR merchants and 'spin doctors' its prime contemporary practitioners?[3]

That is our world now, but it derives from a series of different worlds over centuries. This chapter ranges from the Renaissance to the present, from theological contexts to political, and from the cultural roles of doctors to those of poets. It is intended as a 'background' for the more detailed chapters which follow, and as a way of placing technical developments in wider contexts. It is also a preparation for the final chapter, in which I will discuss the meaning of STM in our times. Some readers may therefore wish to return to it after the rest of the book but before that final chapter.

Let us turn (in)to the magicians of the Renaissance.

Renaissance cosmologies

In some of the literate worlds of Renaissance Europe, it was customary to be joined in meaning with one's surroundings. The world had been created as meaning-full; plants that were good for curing eyes announced the fact by their colours and design. Yet some messages were hidden, and expertise was needed to break the code. The (short) history of the world was written in books, especially the Bible; and from Arabic texts, or Latin, or ancient Greek you could recover accounts of the world that had been lost in later centuries. To read the texts, and to read nature, you needed skills in language and decoding (Debus, 1978; Pumphrey, Rossi and Slawinski 1991).

To enter that mental world, think what it might be like to wake up on a stage, surrounded by a play. How could you work out what was going on and what it all meant? Clearly, you could follow the dramatic action and try to make sense of the plot. You could study the sets and the 'props' for their uses and meanings. Or maybe you could find a text that corresponded to the play, and so discover more about the author and his production; perhaps you could even find ways of communicating with the

author by ritual or prayer. Those indeed were the ways of
alchemy or astrology – elaborate interpretive procedures or calculations, the use of ancient texts, and spiritual preparation and
communion. By making such comparisons we can try to understand the Renaissance view of the world as a drama. Reflexively
and ironically, we can use parts of our own experience to find the
meanings of a (lost?) world, full of clues and meanings.

To understand diseases in such a world, you need to think
about 'agency' as well as nature. This was a world of hidden personal politics and of courtly suspicion. So if you asked your
physician about your ailments or about the best day to go
hunting, he might focus on blind conjunctions of cosmic forces
or on the machinations of malign agents – whether human or
creatures of the dark – any of which might be responsible for disturbing your humours or planting in you the seeds of disease.
Against unfortunate cosmic conjunctions, you might be fore-
warned by astrologers; for protection against the malign, you
could pray, make offerings and take precautions. Such thinking
is not dead, even in the 'advanced West', but it has not been
respectable among the highly educated since the seventeenth
century, when the reputations of 'magic' and astrology declined
steeply, for reasons we shall discuss below.

To recover that Renaissance world is easier these days than it
would have been soon after the Second World War. The 'literary
turn' in history, philosophy and social sciences has familiarised
much of the reading public with perspectives which would have
seemed restricted and remote in 1960. We are self-conscious now
about reading written texts, and some of us have been taught to
'read' rituals or social interactions as if we were anthropologists.
We refuse to be superior to 'primitive mentalities', and we try to
understand sympathetically the different meanings that are read
into the natural world by various cultures, as well as our own.
Indeed, the hold of hermeneutics in some parts of social studies,
including parts of science studies, is now so firm that we can
easily feel the contrary historical problem – how does one 'get
out of' hermeneutics? How does one stop writing only of what
a bit of nature meant to whoever, and start writing directly about

objects beyond texts, presuming a language that others will share (Schneider, 1993)?

Thus we can revivify key questions about the 'scientific revolution'. How were *parts* of the Renaissance world '*disenchanted*'? How were they 'stripped of meaning' so that plants and diseases, and planets and eggs could (sometimes) be *merely* natural – so that they could be described in 'plain prose' that bore no trace of cultural practices or historical derivations? Or perhaps in a universal language which would perfectly transmit the characteristics of the objects or phenomena in view. Or, to put the same question in terms of ways of knowing and of historical displacements: how was a genre of natural history created or re-created in which 'natural objects' could be constituted without their symbolic meanings, so that sometimes, and for some purposes, we could think of them merely as things? And what happened to hermeneutic ways of knowing as a result of such displacements? In the forms of writing which encompassed this new natural history, where were meanings now located?

In the following sections I follow some of these questions, from the Renaissance to the present. Perforce, I concentrate on literate elites debating the boundaries between the realms of meaning and the realms of naturalistic understandings, or between 'culture' and 'nature', but I try to use 'understandings of diseases' to link elite cultures to the everyday. Here I stress again that we all still know enchanted worlds – many groups live in worlds where God's providence still holds sway, or in 'postmodern' worlds re-enchanted by media and commerce.

Disenchantment?

In his book on *The Order of Things*, the French 'archaeologist of thought', Michel Foucault (1970), gave a striking depiction of the world of symbols to which I have crudely alluded in my discussion of the world as 'theatre'. His emphasis was on *language*; in the Renaissance, words and things were on a level. Things were texts, endowed with meaning; words were to be conjured with, themselves full of history and mystery. Foucault's vision was powerful and has proved influential but, of course, expert scholars of the Renaissance and the seventeenth century are bound to insist that neither the Renaissance nor any other period can be fully characterised in terms of one dominant paradigm. Perhaps we generalisers can meet them halfway by shifting from replacement models in which (Foucault's) 'epistemes' or (Kuhn's) paradigms *follow* each other, towards *coexistence* models in which paradigms *run alongside* each other, with different histories and in different political relations. The dominant form can then become less dominant; the once recessive or subordinate can become dominant. That, crudely, is how we explain political change, and to my mind, the model is useful, even for intellectual history. We do not have to choose between drawing out the continuities or sharpening the revolutions.

Foucault's account focused chiefly on those Renaissance understandings that scholars call 'Neoplatonic'. Philosophers who looked back to Plato stressed symbolism and the meanings embedded in the world, and some of them spoke of the 'world soul'; their doing so was threatening both to the Catholic Church and to the newly emergent Protestant churches of northern Europe. But as soon as we say that, we introduce a politics of knowledge and we begin to move from an 'archaeology of thought' to a more complex social and political history – a move on which many historians would insist, especially if we hope to provide *explanations* of ideological restructurings.

We can further recognise that Neoplatonism was not very common, even if we restrict our attention to the literate elites and so ignore the vast majority of the population. Aristotelian traditions were much more important to the Catholic Church and to the universities, many of which were religious foundations. These traditions were not about secret meanings in nature; rather, they had been carefully constructed by medieval theologians to insert the Judaeo-Christian God into a sophisticated natural philosophy derived from the writings of Aristotle. Moreover, there was much in Aristotelian traditions that would count as 'natural history,' and much, too, that we can regard as analytical knowledge, notably the complex astronomy of Ptolemy in which the seemingly erratic movements of the planets were revealed as compounded of various circular motions.

Yet, even as we recognise the pluralism of the culture, the distinctiveness of these naturalistic and analytical traditions, and their continuities with later natural history and analysis, we can also recognise that all these Renaissance traditions stressed recovered texts and ancient wisdoms, and that classification and causal explanations were, by later standards, *subordinated* to views of the world as a single, ordered, meaningful system (Foucault, 1970). In terms of the 'nesting' of knowledge systems, we may say that hermeneutic ways of knowing were usually dominant. Natural history, as we shall see in Chapter 3, was often read as a preparation for natural philosophy, and analytical astronomical calculations were part of the hermeneutic practice of astrology. If we take medicine as an example, Renaissance practices relied on the reading of ancient texts, and dissection was used to illustrate the texts, chiefly to show the purposefulness of the body's structure – 'the uses of the parts'. The body had meaning, revealed chiefly by refined linguistic skills. When cabinets of curiosities were collected by Renaissance princes, or physicians and academics, specimens of nature sat alongside relics and craftworks – as treasures, oddities and symbols or, if you like, as 'conversation pieces'. Wonders were to be talked about (see Findlen, 1994). Hermeneutics was the overarching way of knowing nature, within which others could nest. Man understood himself through reading his place in a meaningful universe.

Thus, in the terms of a displacement model, the 'disenchantment of the world', beloved of many historians, becomes a matter of changing boundaries and relations between different ways of knowing. As we shall see in this chapter and the next, some seventeenth-century 'subcultures', perhaps especially in Protestant countries, began to give priority to 'natural history' – to description rather than meanings. We shall explore some of the causes and consequences of that displacement, but we can note now that it did not produce only a new 'nature'; it also, necessarily, produced new kinds of literature. Devices such as allegory – *pretending* that nature had meaning, were possible only in 'disenchanted' worlds; only to the extent that 'nature' was disenchanted, could it be *re-enchanted* by language.

C. S. Lewis, the historian of literature and apologist for Christianity, described the sixteenth-century world-reading as 'tingling with anthropomorphic life, dancing, ceremonial, a festival, not a machine', in which, 'the "teeming earth" can almost literally be "pinched" with a kind of colic' (Shakespeare's *Henry IV*). And then, Lewis wrote, 'Man with his new powers became rich like Midas but all he touched had now gone cold and dead. This process, slowly working, ensured during the next century the loss of the old mythical imagination: the conceit, and later the personified abstraction, takes its place' (Lewis, 1954: 6). In this view, meaning became separated from 'the natural', and the reading of texts was separated from readings of the world.

Most historians would agree that we see here new boundaries that were formative for later Western cultures. But if we can see such shifts as *dis*placements rather than *re*placements we make more room for pluralism and for ongoing debates and tensions around the boundaries of the natural and the hermeneutic. Two such tensions seem particularly important here. To one of them C. S. Lewis alluded when he suggested that Romanticism in the decades around 1800 tried to heal the pain of what we might call the Midas effect. We shall return to this opposition and tension towards the end of this chapter, to see how Romantics such as the poet William Wordsworth then reinvested nature with moral meanings – a reinvestment which proved fundamental for later cultures, whether religious or secular. The other tension, of course, concerns theology. God remained central to Western intellectual cultures until the twentieth century, and theology was the chief vehicle for debating the relations between nature and men. Those seventeenth-century authors who celebrated such new inventions as printing or gunpowder, found in the disenchantment of nature a new basis for theology, a 'natural theology' of a *God who was also a maker*. This remained a crucial tradition in British and American reflections on science from the seventeenth century onwards, and it echoes on through our own times,[4] even for secularists or the irreligious.

Natural theology and natural diseases

It was, in part, to glorify God as a free creator, that He was freed from his creation. The intermediaries between God and man – the angels and saints, and the demons – were stripped out by Protestant theologians as part of their attack on the old Church, and 'man' was left to face his maker alone, as part of the religious-politics that proclaimed 'the priesthood of all believers'. All good Protestants were assumed capable of interpreting the Bible on their own; and just as the Bible could now be read by anyone, so also the 'Book of Nature' was open to all. In this way, nature became a means of demonstrating the attributes of God through the study of his creation, and religion became a major reason for studying natural history and natural philosophy. In the continuing Catholic traditions, by contrast, life's meaning could more easily be lodged in the Church authorities, in the priests and in the persistent saints. In Chapter 3 and 4 we shall meet some of the new 'naturalists', and we shall flesh out some of the reasons why this new view of the world suited the aristocrats, merchants and craftsmen to whom the Protestant parsons ministered.

In part this was a matter of natural history. They were all interested in selling and owning *things*, and those interests supported the collections and the data-gathering which was sometimes directly useful, but which also helped celebrate man's possessions *and* God's creation. Natural history was part of material culture, but it was also moral; and in as much as it was about hierarchy, it also bolstered the social pyramid. For example, in the years after the Restoration of the English monarchy in 1660, taxonomic hierarchy became a common theme both in natural history and through the cultivation of heraldry.

By including technology, 'natural history' could also exhibit *human creativity*. This, too, had both a mundane and a symbolic

significance – for practical life, but also for religion and philosophy. In stressing the novel creations of man, and especially the
new crafts of printing and gunpowder, the Elizabethan lawyer,
Francis Bacon, had given a new twist to the perceived relation
between man as knower and man as creator. For Bacon and his
followers, man knew *in as much as he could create*; human
power was both the evidence and the fruit of true knowledge
(Pérez-Ramos, 1988; Rossi, 1957; 1970). Here, was a doctrine
which raised the status of 'ordinary life', and of crafts which had
been beneath the dignity of ancient and medieval scholars. It
appealed greatly to puritans, who liked to praise their God in
daily life rather than by special ceremonies. As Charles Taylor
has demonstrated so well, much of the cultural repertoire of
modern man can be traced to this double emergence – to the
inwardness of man the knower, observing nature from the
outside, and to the exaltation of ordinary life (Taylor, 1989).

But the men (and occasional women) who made or supported
the new natural philosophies were also interested in what we
may call 'analytical order' – in astronomy and mechanics, in the
regularities of the world. This interest, yet again, was both
mundane and symbolic. Analytical knowledges and practices
were important for navigation and surveying – for mastering the
divisions of space and of time, but they also helped provide legit-
imations for social order that were independent of Church tradition; the world was regular, its regularities could be grasped,
and they were grounded in God. To seventeenth-century modernists such as the mathematician and philosopher René
Descartes, the ability of man to grasp the elements of nature as
matter in motion, was itself a matter of divine dispensation. He
and other 'mechanists' found models for God's creation in
human creations, especially in clocks (Mayr, 1986); their wonderfully elaborate workings could be understood without calling
on their makers, but their *existence* (and maybe their forms and
varieties) was otherwise incomprehensible. So too with 'de-
souled' nature: it required and reflected its Creator. At *that* level
it evoked the divinity of order.

As Keith Thomas has argued, these shifts in theology and

metaphysics were part of the explanation for the decline in the status of witchcraft and astrology, neither of which fitted the Protestant cosmology in which humans faced an all-powerful divinity (Thomas, 1971). God was not bound in the way that astrology seemed to require, and the puritan world-reading of Man facing God left little room for 'spirits'. What then did this mean for moral responsibilty and for disease-causes other than 'spirits'?

The consequences were twofold. On the one hand, health and disease were matters of natural law that could be studied by naturalists. On the other, God *could* intervene directly; so a sore throat was a natural condition but it might also be divine punishment, and an epidemic could be the judgement of God on the faithlessness of his people. The *private* forms of that perspective on the world shine through the diaries of many devout Protestants from the seventeenth century onwards – their intense concern with bodily regimen *and* with spiritual health of the soul. Such attitudes were often encouraged in the lower classes, especially in times of political instability. In Britain and the USA, they were a major root of the 'self-improvement' cultivated by nineteenth-century workers, and by the twentieth-century's 'positive thinkers'.

The *public* roles of 'divine punishment' were much debated in eighteenth- and nineteenth-century theology. 'Atheists' or 'deists' stressed the mechanisms and the material of nature, so casting God out of the picture, or reducing him to an original cause – sufficient to lend respectability to their own pursuits but too rarefied to serve the purposes of the clergy. But these were not just academic debates, they affected common life. So, for example, changing ideas of health and disease can illustrate the consequences of philosophical differences, and perhaps also help us explain them. The historian Michael Macdonald has provided a nice case study, in which he showed that in early eighteenth-century Britain, the magistrates responsible for local 'law and order' came to reject the tradition that suicide was a heinous sin; they exercised compassion for the relatives by treating suicides as lunatics, and by classing lunacy as a physical disease rather

than possession by spirits. Doctors were happy to agree, and to
assume authority thereby, but the pressure seems to have come
from laypeople weary of the strenuous religious disputes which
had led to the English Civil War. They remained suspicious of
religious zeal as unsettling to society, and were relieved to treat
'enthusiasm' (which meant 'God within'), not as the commend-
able sign of holy possession, but as a rather regrettable lack of
balance (Macdonald, 1990; 1981).

Enlightenment philosophers argued about the possibility of
miracles; deists and 'liberal' Protestants decided against them.
When the 'better classes' of Georgian England set up charity hospitals, though they included the services of clergy, they put the
emphasis on doctors and on the physical recovery of the patients.
(No place here for dying in the sight of the cross, as was still the
case in the great French Catholic poorhouses that we shall visit
in Chapter 5) (Risse, 1999). The French might set reason against
the Church, but Englishmen saw concord. Hygiene, for example,
accorded with the laws of man's nature, with the laws of society,
and with the laws of God. For the enlightened, a cool divinity
underwrote the rational order, both in the heavens that Sir Isaac
Newton had explained for his countrymen, and in the streets
around the assembly rooms that Jane Austen's novels still evoke.
But even then, evangelical Christians looked for personal salva-
tion rather than divine order, and saw God's wrath in public cat-
astrophe. That tendency increased and became more influential
after 1789 when the rulers of all Europe were unsettled by the
French Revolution, and the owners of Britain were worried by
the new self-consciousness of the working-classes.

Revolution, respectability and evolution

After the French Revolution and in face of the unrest of the new
industrial cities, theologies hardened. As natural theology was
recruited to the defence of the Creator, so individual responsibility and hellfire were evoked to preserve moral order (Ward,
1972). In 1832, for example, amid the turmoil of political

reform, the Anglican Church proclaimed that the novel and frightening cholera epidemic was a divine punishment for irreligion, though 'Liberal' theologians still preferred to stress public filth and other 'natural causes' (Morris, 1976.)

British physical 'scientists' busied themselves with analysis and called on the divinity chiefly as the ground of the universe, but the 'fittedness' of animals and plants to their surroundings (and of their surroundings to them) remained crucial for *natural history*: God was the 'great gardener' who brought species into being where they fitted best. Only political radicals were likely to dispense with this explanation of the complexities of life, and they then had to use Lamarckian theories about which the leading biologists were condescending, as we shall see in Chapter 5 (Desmond, 1989). That is why Darwin's *The Origin of Species* (1859) proved so controversial; his doctrine of evolution by natural selection gave a scientifically respectable explanation for the variety and adaptation of organisms, without invoking a creator. For many of those attached to natural theology, Darwin seemed to take the meaning out of life; and in as much as *man* himself might be a product of evolution, Darwin threatened his special status as a moral being made in God's own image. It has been plausibly argued that Darwin allowed about twenty years to elapse between his insight and his book because he was worried sick about the likely public consequences of his views (Desmond and Moore, 1992).

But by the later nineteenth century, when the threat of revolution in Britain had subsided and the middle classes felt more secure, public theology gradually softened and more accommodation was sought, both between the various religious denominations and with the newly professionalising sciences. Bishops might speak for God during epidemics, but state agencies treated individual or collective disease as primarily a matter of science rather than meaning. Public education, in the new state primary schools, or in new universities and technical colleges was non-denominational or even secular. Formal prayers and Bible readings were compulsory for children, but older students were increasingly allowed to separate public learning from their private religious affiliation (or lack thereof). Though most scientists and doctors were Christian, a few prominent scientific educators, e.g. T. H. Huxley and John Tyndall, proclaimed themselves agnostic and battled to ensure that institutions of science were developed separately from those of faith. They drew on Darwinism as science to which religion must accommodate, and on the new sciences of electricity and bacteriology as evidence of science's contribution to human welfare.

Similar shifts in public discussion took place in the USA, especially after the Civil War of the early 1860s. New institutions of secular learning (often derived from Germany) and a new faith in 'professionalism' may have helped bind a riven country and unify a religiously diverse nation. Science was high-learning, but it was also practical and good for economic development.

Generally in the West, the new theories and technology were accepted as evidences of the progress of Christian nations, but ones which could involve readjustment of faith and public morals. For example, when introduced to the world from the 1870s, the bacteria of the French and German scientists disturbed Miss Nightingale and the leaders of the hygiene movement. The microbes seemed too arbitrary in their attacks on humans. When cleanliness had meant health and dirt meant disease, then the science of hygiene had taught morals; but the lesson weakened the more that *anyone* was at risk of bacterial invasion – as often seemed the case when late nineteenth-century medical scientists tried to distance themselves from older styles of public health, and when the lobby for more state science and more science education was establishing its secular authority (MacLeod, 1996; Turner, 1980).

Science, progress and the State

By the end of the nineteenth century, 'science' seemed a motor of 'civilisation'. For some (neo-traditional) Christians, evolution by natural selection had become a substitute for divine providence. Advanced opinion saw evolution and human history as progressive – as a gradual unfolding of creation which could be assisted by a better understanding of biology or of economics. This approach could be further articulated via a naturalistic study of 'morals' and especially via the study of 'moral behaviour' in children. For just as the study of epidemics suggested how you could minimise them by improving public hygiene, so studies of psychology would reveal the human condition and suggest better ways of coexisting peacefully and constructively. Conduct might be understood and guided by new analytical science, such as psychology, craniology and sociology (Smith, 1997). Around 1900, child-rearing was the object of much science, and many white 'progressives' were looking to the new 'genetics' and to 'eugenics' in the hope of improving their (part of the human) race by better breeding (Kevles, 1985).

Such views were common across Europe, even in countries which had experienced rather different relationships between theology, politics and STM. In Catholic France, natural theology had always been much less important than in the UK or the USA; the usual opposition was between the traditional authority of the Church and the progressive, secular, possibly republican culture of the schoolteachers and intellectuals. Evolution was elided with the progressive view of history; the analytical sciences were a basis for the technocratic politics favoured by many scientists and partially realised after the German occupation of 1870, when France sought to catch up with the technology of her stronger neighbour (Fox and Weisz, 1980; Geison, 1984).

By the end of the nineteenth century, the new technologies of electricity and medicine were the material evidence of French progress, though France's medical champion, Louis Pasteur, did not fit the stereotypes at all well. He was a devout Catholic, whose suspicion of scientific materialism had fuelled his experiments to show that fermentation (and disease) were due to 'life' rather than chemicals. But he was also a fervent nationalist and a believer in the utility of science. The agricultural and industrial technologies supported by his work – on wines and beer, and on plant, animal and human diseases – were central to the 'modernisation' of France and its empire. His experiments promised action against disease; the success of immunisation confirmed the potential and made him a national hero. In the cottages of rural France, charms against spirits could now keep germs away (Latour, 1987a; Salomon-Bayet, 1986).

Much of Germany was Protestant, but the 'critical philosophy' which had been part of the university culture from the late eighteenth century saved most German intellectuals from the spiritual travail of British and American believers. 'Culture' rather than God the creator was the first guarantee of social order, and German scientists learned from the philosopher Immanuel Kant how to separate the realms of scientific knowledge from questions of religious belief – they were different ways of knowing: one was about causes, one about purposes and values. True, some mid-century radicals tried to make scientific materialism into progressive politics by claiming that (analytical) biochemistry was the sure way to understand the mind, and that politics could be part of medicine, but later nineteenth-century STM was generally more conservative. Scientific industry, scientific medicine and scientific warfare became strongly identified with the imperial state.

This was true of the new industries, which the emperor backed even when professors were stand-offish; and it was true for the new bacteriology of Robert Koch, which was closely associated with the military – in marked contrast with earlier 'public health' which had been liberal in its politics and based on British models (Evans, 1987; Johnson, 1990; McClelland, 1980). The new, continental, bacteriology was not about better drains or nutrition; it focused on specific diseases, technical fixes and regulations against contagion. It was a very analytical, bureaucratic *Prussian* science, but one which still echoes as part of our repertoire when thinking about public health. To take a recent example – in the early years of the AIDS epidemic in Britain, there were proposals for the compulsory isolation of some cases, and historians were mobilised to discuss the precedents (Evans, 1987).

Generally, by the opening of the twentieth century, European and American STM was increasingly associated with the drive for national efficiency in a world of industrial and imperial competition. In some ways, the horrors of industrialised warfare between 1914 and 1918 were but the tragic epitome of this modernist mobilisation. Similar arguments might have been made for the Soviet Union under Stalin, and for the National Socialist government of Germany after 1933 which allied technocratic efficiency with atavistic politics. And as we shall see in the final chapter, critics of American science in the 1960s saw government-supported STM as bound into military-industrial and medical-industrial complexes which they found alien to the liberal politics and free markets supposedly characteristic of America. Indeed, throughout the twentieth century, with more or less reason, the threat of the technocratic state was part of the public meaning of STM. But by the end of the century, it was the expanding alliance of new STM and global capital which had come to seem most problematic.

Modernist human-natures

Very generally, the more that twentieth-century science was presented as offering new technologies, remedies and protections, and the more it was identified with 'modernisation', the looser its articulation with 'traditional' society and morality. Early twentieth-century modernists argued for lifestyles that were 'scientifically natural', based on biology, informed about sunshine, open about sex (Porter, 2000). In some ways, they naturalised the hygienic rules (and the streamlined, dust-free architecture) that could be found in mountain sanatoria for the tuberculous. Such 'progressive' lives *could* be religious; they might be 'Romantic' in their attachment to nature, as we shall see; but the guiding principle seemed to be the fulfilment of human capacities by rational means.

Looking back, the 1950s now seem the high point of our collective confidence in the ability of technoscience and of STM more generally to protect us from danger and to steadily raise the 'standard of living'. Those of us who grew up under the British National Health Service, drank our orange juice and gulped

down our cod liver oil, and if we were still puny we were taken
to the school clinic in town, where a rotating 'sunray' lamp
spread health around the darkened room. For a quarter-century
or so, post-Second World War, antibiotics and immunisation
seemed capable of subduing infections, nuclear energy of providing clean power, and 'development' and contraception of
raising the Third World. Because some diseases could be stopped
by antibiotics, bacteria became their one and only 'cause'; the
other preconditions that had worried hygienists seemed less relevant now that a cure was to hand. For a new generation, acute
diseases and medicines became less 'biographical', less a matter
of morals or even of lifestyles (Bud, 1993; 1998). It was only for
the emergent chronic diseases of older people, that biography
mattered; for the rest, diseases could be conquered and progress
seemed assured – except for the threat of nuclear war.

We are less sanguine in 2000, but still the rational progress of
medicine is part of our common culture. While acknowledging
the links of new infections with lifestyle, our public culture does
not see disease as God-given. Though 'fundamentalists' of
various kinds read the AIDS epidemic as the *expected* result of
moral transgression (as Miss Nightingale would have agreed),
public health agencies see an *unexpected* infection which, in the
West, has afflicted groups that were atypical, but neither 'abnor-
mal' nor immoral. In this account disease is a risk, but not a
morality play (Berridge, 2000).

Yet these histories are complex and ever changing. One could
say, for example, that in the 1950s homosexuality was a treat-
able disease; for radical counter-culturalists of the 1970s it was
a nature that should be expressed; and now, perhaps, a set of
inclinations which the wise will recognise in themselves and in
others, but which *require* neither suppression nor expression. Yet
one could also say that all those attitudes have been variously in
play throughout the last half-century. We could trace the conflicts between their proponents, and consider the ways in which
the 'balances of power' depend on formal politics, and how they
vary between nations and regions.

But in doing so, we would have to recognise that we are not

the only historians; the people we study see themselves in history. They present themselves as progressives, or as custodians of traditions which have withstood repeated attacks; they draw on formal and informal histories to constitute themselves and their campaigns. More than 'nature's laws' are at stake in determining appropriate responses, and even if we focus on the interpretations of nature, modern STM is not the only authority. Present ideas of what is 'natural' for bodies and for environments are often rooted in *reactions against official STM*; they may be as much the creations of popular Romanticism as of popularised science (Berridge, 1996; Shilts, 1987).

When atheists dispensed with God in the late nineteenth or early twentieth century, some thought that the human sciences would be sufficient guides for conduct and politics. But this was a rare position, for then as now, naturalistic accounts of man were not the only alternative to formal religion (Lepenies, 1985); the human condition and its meanings could be explored through culture, through the hermeneutic worlds of philosophy and literature, and by seeing in 'nature' a source of inspiration rather than information. To explore that strand of meaning, we conclude the chapter by returning again to 1800 and to the reactions against analytical science.

Nature and culture

Some nineteenth-century novelists regarded themselves as natural historians, as analysts of humanity or even as experimentalists, but other Victorian traditions seemed more at odds with 'science'. For example in Matthew Arnold's *Culture and Anarchy*, we see a familiar formalisation of 'two cultures': the opposition between the sciences that told you how to do worldly things, and the humanities through which you discovered what to do and why (Williams, 1958). This opposition derived from Germany, it was characteristic of many brands of idealistic philosophy, and it remains crucial to our intellectual world. But it is important to note that the opposition was and is chiefly between 'approaches' rather than the topics addressed. Both nature and human history could be read for facts and laws *or* for feeling and values; or, indeed, through the various combinations of these approaches which remained widespread in Western culture throughout the twentieth century.

In large measure, this opposition was a product of Romanticism, a complex set of reactions against the eighteenth-century rationalism which had pictured man as building up knowledge like a naturalist arranged plants, or as an analytical machine reducing complex phenomena to simple principles. The very success of 'analysis' in the decades around 1800 contributed to the Romantic reaction against 'dissection' and reductive explanations – though new types of 'analysis', notably morphology, could be thought of as anti-reductive (Cunningham and Jardine, 1990). Romanticism is indeed complex, but one key lies in the presumed relation of nature and man, described by Charles Taylor as follows: 'The meaning that natural phenomena bear is no longer defined by the order of nature itself or by the Ideas which they embody. It is defined through the effects of the phenomena on us, in the reactions they awaken. The affinity between nature and ourselves is now mediated not by an objective rational order but by the way that nature resonates in us' (Taylor, 1989: 299).

To take a widely known and influential example: in the philosophical poems of the English Romantic, William Wordsworth, the sense of nature as a teacher of morals is overwhelming. External nature resonates with the soul that is the nature in us at our birth, and for us to contemplate the hills and plants, or the lives of the people close to nature, is to cultivate one's moral self (Gill, 1989). In the later nineteenth century, John Ruskin, the art critic, painter and amateur geologist, developed these same sentiments and linked them to trenchant attacks on the ugliness of capitalism. At one level, theirs were powerful calls to the cultivation of a kind of 'natural history', and such sentiments have remained a major motivation of naturalists and conservationists – from the English Lake District which was home to both Wordsworth and Ruskin, to the national parks of America, and now to global nature-tourism (Gould, 1988; MacKenzie, 1988). But Wordsworth (and Ruskin) drew the line at reductive analysis – 'we murder to dissect' – and that attitude, too, has echoed to our time.

We see it clearly in some of the feminist critiques of science from the mid-nineteenth century onwards, most literally in the anti-vivisection movement that first thrived in England around 1870, not least among women (French, 1975). We shall note in Chapter 5 that certain kinds of nineteenth-century science could be termed Romantic in their concern with unity and development rather than analysis into components. The ongoing debates around feminism and science pose questions about the plausibility and productivity of 'alternative' styles of science and about approaches to biology that would be 'more respectful' of organisms as wholes (Harding, 1991; Harding and O'Barr, 1987; Keller, 1983); and that same orientation is also evident in popular literature on the 'holistic' approaches to nature preferred by 'ecologists'and by critics of 'reductionist' medicine (Bramwell, 1989; Worster, 1994). Behind all these complex and political debates, we see the tensions between nature as the object of classification and analysis, and nature as 'whole' and meaningful; we see the boundaries moving and blurred, always in dispute. Sometimes natural-historical knowings were contrasted with analytical methods, and sometimes 'natural history' was itself contested – between those who stressed classification and order and those who stressed meanings and individuals.

And this was also the case for human history – was it an extension of natural history (or even a field for analysis), was it to be searched for evidence of God's providence or for laws, or was it to be contemplated as a means of understanding other ages 'from the inside'? To what extent could these goals be compatible? These questions were crucial for nineteenth-century configurations of science and scholarship; they remain important for us (Bowler, 1989b; Levine, 1986; Mandelbaum, 1971; Outhwaite, 1975).

History here, of course, includes the history of STM, and the texts that can be 'read' hermeneutically include the papers and books of past or present 'scientists'. *If* these are read primarily for the construction of meanings, then they become part of a hermeneutic history. In such ways the 'nature' presented by STM becomes a reflection, not just of God or of how *things* really are, but of the investigators who produced the texts, and of the cultures which shaped them. R. G. Collingwood, who had read 'old' science books as a child in John Ruskin's library, realised that history could be regarded as prior to the 'construction' of nature. Recent history of STM follows this insight, to discover how all knowledges have been created within systems of meaning over time. So here we return to the themes of the start of this chapter; here, indeed, is the place of this book in my account of Western culture. I use historical analysis to reconstruct the many 'projects' and 'understandings' in the past and present of STM; history becomes part of our reflections on nature and science.

With that circle completed, we turn now to focus on the histories of the ways of knowing that comprise modern STM. We turn from the writing of human history to the making of natural history – and then to analysis, experiment and technoscience – but always trying to see the projects that constituted modern STM as also constitutive of our human histories.

Notes

1 But I do not mean to suggest the opposite mistake – an eternal 'natural-historical' language for patients and symptoms. Clearly, some aspects of technical medicine are domesticated in patients' lives as drugs and/or as concepts. Just how they are understood and used is not well recorded – we need more studies of the hermeneutics and natural history of life under technoscience. See Cornwell (1984) for one model, or chapter 7 of Collins and Pinch (1998). Perhaps some of the tools developed by historical sociologists for analysis of interactions between professional groups could be applied here. See, for example, Star and Griesmer (1989) on boundary objects, or Peter Galison (1997) on 'creole' languages in physics. See also Pickstone (1994b) for the various material cultures of medicine, and Miller (1991) for modern material cultures more generally.

2 To see how good that teacher is, read Bray (1997).
3 See also Gouk (1997).
4 See the two excellent surveys of science and religion by Barbour (1966) and Brooke (1991).

3

Natural history

THIS CHAPTER IS ABOUT KNOWING the variety of the world – about describing and collecting, identifying and classifying, utilising and displaying; it is about the 'notebook' cultures of men and women who loved to 'take note' of their surroundings – not chiefly for meaning, nor necessarily for use, but for the wonder of it or from a compulsion to identify and collect. As already discussed, the term 'natural history' here covers all the things that can be named and collected, not just the animals and plants to which 'naturalists' now are usually devoted. The chapter analyses the history of collecting, describing and displaying – for pride of possession, for intellectual satisfaction, and for commerce and industry.

Natural history was a common term in the seventeenth century, when it had the wide meaning I use in this book. It was the register of facts, the compilation of what was in the world. It contrasted with natural philosophy, which was the account of *causes,* a matter of explanation rather than inventory. Natural philosophy will be discussed further in the next chapter; here I concentrate on how inventories became possible and desirable, and the forms they have taken. We have already seen how natural history could be set against hermeneutics, and 'natural objects' might be 'revealed' by 'removing' the symbolic meanings, the mythology, etymology, etc. But that formulation is problematic; it may suggest that what was removed was 'accretions' whereas it was in fact an integral part of the meaning of a plant or an animal in the Renaissance cosmos. So rather than asking how meanings were stripped away, we do better to ask

how 'natural objects' were *created* in the move away from
Renaissance worlds.

The tradition of natural objects as bearers of meaning had been
nicely epitomised by the cabinets of curiosities which became
fashionable from the sixteenth century, and which have recently
attracted the attention of historians. The 'cabinets' were meant to
be amazing; the collections were of many and varied objects –
found or made, each with its own 'story' or message. A horn from
a unicorn (or narwhal), a monstrous newborn cat, a stone with
the shape of a fish, a rare crystal or jewel, antique coins, or
perhaps stone implements; they were objects of significance and
of prestige (Findlen, 1994).[1] Such collections were often maintained in high-status sites through to the later eighteenth century;
at popular level they continue to our day, in private houses or in
'freak shows' (Altick, 1978; Elsner and Cardinal, 1994). They are
entertainments, to be sure, but we do well to remember *that*
aspect of *all* displays; it is not by the appeal of zoological taxonomy that the great natural history museums attract so many visitors; and no matter how 'scientific' the exhibits, most visitors
will latch on to oddities and remember the museum for the great
spindly spider crab that sat in a box at the top of the stairs. Our
historical problem is not to explain the continuation of that
hermeneutic aspect of collections, but to see why it was partly displaced by more sober ranks of specimens which were meant to be
seen, not so much as individually meaningful, but as parts of the
ordered arrays of God's creation and of human artifice. Why, by
the eighteenth century, were cabinets of curiosities replaced by
new collections prized for their 'coverage' of plants or shells, minerals or coins (Hooper-Greenhill, 1992)?

As always in explaining such developments historically, we
look for (and to) alternative approaches that could be drawn on
and developed – perhaps for subsidiary aspects of culture that
might become more prominent; and we look for 'contexts' in
which such developments seemed desirable. Here we note two
such approaches, which we can call Aristotelianism and the new
art; and two such contexts – the exploration of new continents,
and the celebration of earthly life and its material riches.

'Historia' and representation

Within Aristotelianism, classification had been a major concern, and so important was this tradition that 'natural history' as it emerged in the seventeenth century might be regarded as a modification of Aristotelianism rather than an alternative to it. The academic botanists of the Renaissance often saw themselves as continuing the work of Theophrastus, the disciple of Aristotle who had focused on plants. Aristotle indeed remained a major resource for 'biologists' well into the nineteenth century; Charles Darwin spoke of Linnaeus and Cuvier as his masters – but 'they were mere pygmies to old Aristotle'. If historians of science tend to disparage the easy naturalism and 'purposiveness' of Aristotle's universe, it is partly because they overlook questions of description and classification in everyday knowledges.

We have argued above that Renaissance scholars tended to subordinate description to meaning. Texts were crucial and natural history was subordinated to natural philosophy. If animal and plant forms were described, it was primarily to show the ways in which the parts functioned in support of the whole, and the ways in which animals were 'self-crafting' – realising in development the purposes inherent in the 'seeds'. Gianna Pomata has argued that part of the seventeenth-century shift to a new form of natural history was an upgrading of description, a readiness to *record* while worrying less about the underlying principles. Doctors, for example, began to publish case histories, as specimens of disease and of their practices. Plants were described 'for their own sakes', not just as remedies (Pomata, 1996).[2] We may link these changes of attitude to the discussions in the previous chapter about market cultures and the celebration of everyday life, about the growing appreciation of crafts and the perception of God as craftsman rather than 'author'. But we can also link them to new techniques of visual representation.

If Aristotelian naturalism was reborn in the universities, art was reborn in the cities of Renaissance Italy, especially in fifteenth-century Florence. It was, and remains, a much more conspicuous feature of their cultural achievement, but one which we as heirs of Romanticism too easily place in opposition to 'science'. We may need to be reminded that the new architecture and naturalistic representation of human bodies were intellectual as well as aesthetic ventures. Central to these arts and sciences was the creation of *perspective*. As Santillana claimed, the 'discovery of perspective, and the related methods of drawing three-dimensional objects to scale, were as necessary for the development of the "descriptive" sciences in a pre-Galilean period as were the telescope and the microscope in the next centuries, and as in photography today' (Santillana, 1959: 33).

From Brunelleschi's experiments in the early fifteenth century, to Masaccio's frescos in the Santa Maria Novella, architects and artists, especially in Florence, struggled to show spatial relations on a plane as they appeared to the eye, e.g. in a camera obscura. In depicting buildings or bodies, the parts were to be related to each other in this perspectival frame. These spatial relationships were central to the new anatomy, best known from the magnificently illustrated volume, the *Fabrica*, published by Vesalius in 1543. As scholarly anatomists pored over editions of Greek texts to present a reborn account of the human body, so artists explored the relationships of the parts and found ways of depicting 'innards' that were as 'lifelike' as the external features more commonly presented in paintings and sculptures (Kemp, 1990; 1997).

One can see it happening. In medieval anatomy texts, all the illustrations are diagrammatic; in Vesalius, all is naturalistic – bodies stripped of skin, or variously splayed out, or skeletons posed in landscapes. In between, *c*.1500, you can find naturalistic bodies with diagrammatic innards – Renaissance outsides, medieval insides. The new pictorial discipline, presenting the world as to the eye, helped open the human body as a land to be explored (Herrlinger, 1970). Where medieval anatomies had been demonstrations of theory, by the seventeenth century they were often seen as part of the new natural history.

Throughout the early modern period, *exploration* was a key to anatomy, and parts of the body were 'seen as' other parts of the world – as fountains and streams, or as branched like plants, etc. The new anatomy paralleled the voyages of exploration then revealing the contents of new continents and shipping them back to the old.

New worlds, new properties and new creators

The exploration of lands unknown to the ancients, and the enormous collections of new plants and animals built up in the capitals of the trading nations, required new forms of inventory. But specimens from the New World came to Europe without the mythology and emblematic significances which were part of the classical tradition for European plants (at least for medicinal and culinary herbs). It is an instructive coincidence that the first botanical texts to include New World plants were also the first to drop the 'human', symbolic aspects, to create natural histories containing 'nothing of the Imagination' (Ashworth, 1990: esp. 322; Grafton, 1992). Likewise for archaeological specimens – remains found in Northern Europe, like New World plants, did not easily fit into written histories derived from ancient Greece and Rome. For Roman coins there was a rich literary context; much less was available for British remains. So these too came to be treated as physical objects which could be listed and catalogued, to form part of a new natural history.

But *why* were these new kinds of materials being imported and/or collected? What created the need for taxonomy? Here we turn to the growth of urban commercial society (especially in Holland), but also to religion (and especially Protestantism, as introduced in the last chapter). We begin our exploration with notes on seventeenth-century Holland, and then move to eighteenth-century England, before returning to questions about the knowledge forms of natural history.

The Dutch case has been beautifully analysed by the art historian Svetlana Alpers in her pioneering book on *The Art of Describing* (Alpers, 1983). She stepped outside a tradition of art history rooted in classical, heroic, narrative art which was meant to be 'read', and she explained Dutch art as celebrating objects and possession. Her critics have argued that she underestimated the narrative, historical and moralising aspect of Dutch painting, but at least, Alpers's shift of emphasis challenges us to analyse forms of depiction and description that often seem so natural (to us) as not to require explanation. (This chapter seeks to do the same for natural history.) Alpers evoked a prosperous urban society, devoted to the arts of peace rather than war, to trade and exploration, and rather liberal in its Protestantism. Rich merchants collected natural objects along with works of craft and art. Paintings were 'pieces of art', well-crafted images that might 'stand in' for the originals (e.g. flower paintings). 'If the theatre was the arena in which the England of Elizabeth most fully represented itself to itself, images played that role for the Dutch' (Alpers, 1983: xxv). These images were 'naturalistic' more than hermeneutic, and the images were everywhere – printed in books, woven into tapestries or linen, painted on to tiles, and framed on walls. Everything was pictured – from insects and flowers to Brazilian natives and the domestic arrangements of Amsterdammers. Maps were central to this culture, not just as tools for navigation, but as representations of the world. The atlas was a Dutch specialism, and so was topography – the art that linked maps with landscapes.

We have already mentioned the crafts of representation in Renaissance Italy; the Dutch could also draw on lenses and the camera obscura – the 'natural' way to produce an image of nature. The image of a town projected in such a camera became part of the world it depicted. The new telescopes and microscopes projected the very distant or the very small so they could be captured in drawings – by 'a Sincere Hand and a Faithful Eye' (Alpers, 1983: 72–3).[3] Clarity and faithfulness of vision was to be treasured, not least by those who needed and could afford the new spectacles. It was in Holland that lens-grinding became a major craft.

Much of the recent writing on the 'scientific revolution' acknowledges Alpers's historical insight. The best recent studies on experimentation in the seventeenth century make clear the relation of these novel manipulations to the wider culture of 'facts and objects' which contemporaries called natural history. Shapin and Schaffer, in their classic study of Robert Boyle and the air pump, see scientific apparatus as making visible the invisible. The pump experiments, to which we return in Chapter 6, manifested the 'spring of the air', rather as telescopes showed the moons of Jupiter or the microscope the 'cells' that made up cork (Shapin and Schaffer, 1985; Shapin, 1994).

But the representation of nature was not only a pleasurable interest and a mark of affluence – it could also be a religious act. 'Anatomising' whether on corpses or on microscopical insects displayed the workmanship of God. This 'moral enquiry' was peculiarly evident in the public dissections of executed criminals, which in Holland were often held in churches. These fashionable events were steeped in moralism: through the corpse of a sinner, the anatomist celebrated God's wondrous designs. Though that sentiment was not new, it took new forms and had a new relation to human creation (Schupbach, 1982).

In most Renaissance world-readings, the craft products of corrupt humans could not compare with products of God, but the Dutch bourgeoisie made the comparison, and their readiness to see God as a craftsman is a key to the possibility and popularity both of 'mechanical philosophy' and the new natural history. We are often reminded that mechanical philosophers liked to see organisms as 'mechanisms' – but that was mostly intellectual fancy. We do better to focus on the ways in which the whole known world could now be used in understanding its obscure realms – whether microscopic, distant or newly discovered. In the new natural history, natural objects and craft objects were on a par – items in an inventory which provided endless scope for analogy. This is crucial, for as historians of art have taught us, all seeing is 'seeing as'. The anatomists who discovered lymphatic vessels saw them as rivulets of water in the landscape of the body, and the form of blood vessels in humans resembled those in plants (and vice versa). You could inject blood vessels with wax and strip away the surrounding tissues, to produce a miniature bush – which might be instructive, or merely ornamental. Indeed, such wax creations were sometimes combined with other bits of anatomy to produce a kind of floral arrangement with moral overtones (Cook, 1996). They celebrated the resemblances of animal and plant parts, and the relationship of nature with wax-modelling. We might call them 'mixed media' and think of Damien Hirst's artwork – the sheep in formalin.

In this culture of possession, natural history was necessarily a matter for the craftsmen and tradesmen who supplied that culture, but so it was also for surgeons, apothecaries and physicians (the university graduates of the medical world). If we skim through Dutch collections of objects, we find the skeletons and monsters for moralising and curiosity; but we also find collections of plant material ranging from the fashionable tulips to exotic spices and herbs, from new kinds of wood for furniture to medicinal plants – perhaps stored in the Delft pottery jars which became symbolic of elite apothecaries. Their shops looked like museums; catalogues of creations became continuous with catalogues of commerce. And by the later seventeenth century, the same was true of London.

When the Royal Society was established there after the Civil War it created a museum, including artefacts, many of them exotic. Like much else in the Society's foundation, the museum was intended to advance the useful arts (Arnold, 1992). In this respect the Society (for a while) continued the endeavour of the puritan Baconian reformers who had pursued the improvement of trades, agriculture and medicine as part of their commitment to a New Jerusalem here on earth. 'Natural history', in my sense and theirs, was central to that enterprise (Webster, 1975); and so were 'ways of doing' in crafts and trades, including 'recipes' – then a common item of cultural and commercial exchange, ranging from cooking to skilled trades, from remedies to tips on the investigation of nature. They stand in a tradition that runs from the 'natural magic' of the Renaissance to the household and trade manuals of the nineteenth century, and to the countless 'how to do it' books (and television programmes) of our day.

Though we know much less about medical practices and commerce than about treatises on physiology, medicine seems to fit this pattern of natural history. The education of apothecaries was central to the development of field botany: apprentices were taken on expeditions to collect 'simples', and for some of them the identification of plants became an end in itself (Allen, 1976). Botanical gardens developed out of herbal gardens, set up by universities or by medical guilds. Chemistry, too, was in part a matter of developing new medicines, in part a matter of improving other trades and crafts, including agriculture. By 1700 in Holland, and by 1750 in Edinburgh, much of this new chemistry, botany and anatomy was included in the education of physicians; the knowledge helped your career as a teacher.

Studies of diseases, too, became more 'natural historical'. 'Case histories' were compiled as contributions to the collective store of observations (and as advertisements of the experience, acuity and effectiveness of particular physicians), and some physicians attempted to characterise epidemics by describing the 'constitution' of the place and time, rather than focusing on the constitutions of the individuals involved. This kind of medicine drew on classical traditions, especially the writings attributed to Hippocrates. Its chief British representative was Thomas Sydenham, a reformer of medicine who moved in the same circles as the physician and philosopher John Locke, and who shared that interest in plants and classification which became seminal for so many eighteenth-century 'savants' (Bynum, 1993; Raven, 1968).

Natures for pedigree people

If seventeenth-century Holland serves as the model for the arts of describing and of making in urban societies, eighteenth-century England serves as a model for the rural; indeed, it may well be our best example of a culture in which the natural-historical was hegemonic. It is no accident that this text is moving from Holland to Britain – so did economic and maritime dominance, and so, later, did the centre of Western medical innovation. From the middle of the eighteenth century, the Scottish universities were notable centres of educated culture – admired by radicals in France, influential in the colonies of North America, and linked by Hanoverian connections to some of the German states and Low Countries. In the Scottish Lowlands and the counties of England, the aristocracy and their imitators among the gentry and merchants built a high culture based on country estates – modelled on classical Greece but revelling in modern knowledge and utilities. Estates were 'improved' to look more like classical landscapes – all the better to be painted. Mountains which had once seemed threatening came to be seen as picturesque or sublime; they were visited, pictured and cele-brated in verse. Nature was domesticated and improved, as were fields, plants and farm animals. Cultivated gentlemen collected plants and animals, eggs and snail-shells, minerals and archaeo-logical remains, as well as coins and prints, sculptures and paint-ings. Such collections advertised wealth, intellect and order; and in as much as they commanded respect, they helped naturalise the social hierarchy (Ritvo, 1987; Thomas, 1983).

Amusingly, British 'history' was then rendered visual and col-lectable in Granger's system of portraits of notables – a grid of hierarchy against time, a matrix which could be filled with com-mercial prints of aristocrats, rather as later generations would collect stamps for their albums (Pointon, 1993). Scrapbooks were common and some families bought printed books that could be taken apart so that extra prints (or even specimens) could be inserted. The growing market in descriptive books and engravings mirrored and complemented the collection of speci-mens and of experiences. Works of God, of man and of art were all superposable – such arrangements might be seen as the dilet-tante version of the classical episteme, the grid of knowledge described by Foucault. It was in such cultures that formal tax-onomies, especially of plants, first came into common use. The debates around them are still instructive, for classification remains a fundamental aspect of old and new knowledges, not least in our 'age of information'.

As eighteenth-century philosophers showed, there were two ways to approach classification. One was to group species with many similar features, and so construct a nesting of similarities. All plants that looked rather like roses could be grouped together as a family; within that large group were smaller ones – for those like strawberries and those like apples, and for those even more like roses. And, at a higher level, the big 'rose family' could be aligned with the daisy family or the umbrella-flowered plants, as opposed to grasses, say. That goal, of a natural system, has remained as a recurrent principle of classifiers. In the twentieth century some taxonomists practise a form of classification called *cladistics*; they use computers to try to establish hierarchies based on maximal numbers of 'characteristics' – all treated as equally important. Some classifications of bacteria work on this basis, which some taxonomists see as uniquely 'objective'. For eighteenth-century philosophers, such a method approximated the mental processes of 'association' which were fundamental to learning (Daudin, 1926; Jacob, 1988).[4]

But such systems were impracticable as guides to identification and arrangement because they depended on many characteristics and were subject to perpetual adjustment as new species were discovered. The alternative was to start with simple characteristics and arrange plants accordingly. Around 1730 the Swede, Carl Linnaeus, devised a wonderfully workable arrangement and procedure. In Holland, at the centre of botanical importing, he came to appreciate the need for classifications that would allow botanists, horticulturalists or apothecaries to know whether or not they were dealing with the same species, whatever the geographical origin of the specimen. To this end Linnaeus invented the binomial naming system and made it stick. (So the common daisy still has a generic and a specific name he gave – *Bellis perennis*.) He linked this mode of naming with his 'sexual system' of plant classification, based on the disturbing, even titillating, discovery that flowers had sexual parts (Frängsmyr, 1984; Stafleu, 1971). In the Linnean classification, the number of male-bits and female-bits found together in flowers served as a God-given principle of order, a divine arrangement which proved its utility and enhanced the pursuit of botany, and of those other realms of nature to which Linnaeus devoted his magnificently orderly brain. It was in the patrician culture of Britain that his legacy was most fully developed and a national botanical society established in his name.

We need to recall the excitement of classification as central to knowing. Classification posed the question of man's place in nature, and his relation to other animals. Classification could be the key to medicine: if only diseases could be properly classified we could more easily recognise them and their resemblances. But here, as elsewhere, intellectual aims could not always be realised; it proved impossible to agree on classifications of disease, mainly because there was no agreement about what constituted a 'disease'. If you examined the relevant text by William Cullen, Scotland's leading physician c.1750, the diseases at the twig points of the classification would seem to you but symptoms, for the most part (Cullen, 1816).

But even where one could not achieve a stable classification, one could still describe and compile. The eighteenth century was also the time of encyclopedias, and such collections included technologies and crafts in all their wonderful geographical variety (Gillispie, 1972). One benefit of including artefacts as part of our 'extended natural history' is that we, like the authors then, can see the similarities between craft products and natural species.

Craft products were regional and in some ways they helped to define and characterise the regions. Different regions made different kinds of spades or cloths or houses; indeed, the peasants themselves differed between regions (unlike the 'cosmopolitan' bougeoisie who observed them).[5] Craft and agricultural products were often studied along with animals and plants, minerals and archaeological remains, diseases and the weather – as part of *chorography,* the natural history of *places over time* (Jankovic, 2000). Crafts 'sprang from the soil' like characteristic plants or the vernacular architecture; like plants they were seasonal and affected by the weather. Though we sometimes think of crafts as mechanical, and opposed to biology, this view hardly bears examination. Most crafts converted animal and plant materials into food, clothes, housing or implements, so it is hardly surprising that the products were described and classified like plants and animals.

But we can push the point further, by reflecting on our present knowledges in ways that help us see the past. We now think of many consumer items in terms of 'species' – or 'brands'. Luxury goods – hand crafted, more or less traditional – seem to be understood on 'species' models, while utilitarian (often analytical) characterisations serve for workaday products. We may specify the useful qualities we want from a heating system or an office carpet, but for fine wines or cheese, stylish clothes or racehorses, 'we' demand a quality brand or a prestige breed (Pickstone, 1997). Such possessions were and are the hallmark of social superiority – pedigree breeds for pedigree people.

And if we want clues as to how an eighteenth-century physician could have assessed his patients, then we could draw on the idea of *connoisseurship*, which we understand best from luxury goods or from works of art. Connoisseurs of wines recognise a kind of wine, its origin, vintage and its normal variations; they also know its characteristic faults and 'diseases'. Likewise, to know a patient well was to understand the character, the context, the way of life, the potential and the failings of a person of that kind. Such was the essence of biographical medicine in the classical tradition which saw disease as disturbance of that which was natural to the individual life. To know a patient, as say sanguine or choleric, was to understand his or her natural form of life, and thus to recognise the departure from that nature which a fever, say, might constitute. To treat was to restore the natural condition, by rebalancing the body and its humours, perhaps by bleeding or a change of 'regimen'.

The biographical physician knew to what diseases the patient's nature was liable, and which of these defects might be associated with particular seasons or places. Or to reverse the matter – he also knew the *nature of a place and/or a season*, and what diseases were so associated. Patients had 'constitutions,' but so did seasons and places; medical connoisseurs understood them, as perhaps they still do. However, that natural history of particular places and times could sometimes be 'in tension' with the kind of natural history which worked with 'universal' taxonomies, e.g. of all flowering plants. The former study was central for country parsons and doctors who *lived* the natural history of their own localities, the latter for cosmopolitan travellers and curators who travelled to collect. From the later eighteenth century, plant collecting became a minor profession, part of scientific imperialism.

Natural empires

I have concentrated on the eighteenth century because natural historical understanding was then hegemonic in European cultures of STM. I shall argue in the next chapter that the period 1780–1870 might be called the 'Age of Analysis', because many new disciplines were then constructed which 'went below' the corresponding parts of natural history by 'taking objects apart'. But this does not mean that natural history disappeared, or even that it was weakened. The nineteenth century, one might say, was the great age of 'scientific' museums, a fact which is hidden from many historians of science by their fondness for 'laboratories'. Those who know Cambridge and Oxford may recall that when these country universities first invested in science, in the mid-nineteenth century, it was by building museums, chiefly for natural history (a kind of science compatible with gentlemanly breadth and the liberal arts). Indeed, most nineteenth-century universities maintained museums as part of their teaching and research. It was only in the late twentieth century that some university museums came to be seen chiefly as 'visitor attractions'.

But the greatest investments in museums were by the nineteenth-century nation-states, in their capitals, as expressions of national and imperial power. Paris led the way, when after the Revolution, the former Royal Botanical Garden was developed as a national Museum of Natural History, with professors as curators. Many other former royal collections were turned to public use in comparable ways, including the art collections of the Louvre, and the technical exhibits collected in the

Conservatory of Arts and Manufactures. Under Napoleon, the conquered nations of Europe and North Africa were systematically looted in order to extend the imperial hoards. These collections not only displayed the power of France; by claiming to be definitive collections of the natural order and of great art, they also identified France with science and civilisation. Patronage of museums became a responsibility and a resource of the modern state, and a few devotees gained careers as curator/professors (Hooper-Greenhill, 1992; Pickstone, 1994a).

The British Museum, based on the eighteenth-century collections of Hans Sloane played a similar role in London, along with the botanical gardens at Kew, initially developed as a royal estate. By the mid-nineteenth century, they were seen as collecting places for imperial treasures and as inventories of imperial possessions and resources. The natural history collections were given a huge new building in South Kensington in 1881, well before the government invested in major laboratories. The collections of the (national) geology museum were also moved to South Kensington, alongside the collections of applied arts and of machines and scientific instruments that we will discuss in the next section (Forgan, 1994; Stearn, 1981). Much of the natural history material came from the British imperial possessions. Some of it was collected on government-sponsored expeditions, the most famous being the voyage of HMS *Beagle* in the early 1830s, on which Charles Darwin was the gentleman naturalist. Often, as in this case, the chief purpose was surveying for the improvement of navigation; in mid-century the British navy was unchallenged, so 'research' was a useful stimulus for bright young officers.

Until the end of the century, American expeditions and collections were 'internal' rather than overseas, but it is hard to over-estimate the importance of collections, museums and documentation in the formation of the 'scientific-nation'. The most obvious institutions were the national museums in Washington, and the associated agencies for investigating the geology, botany, zoology and ethnology of the continent; but the commercial and industrial cities also had prestige museums, often, as in Britain, established by societies of enthusiasts. The universities developed in the last third of the century built their own collections, not least for teaching agriculture (Dupree, 1957).

From about 1900, all the major European powers (and the USA) invested heavily in colonial development to improve tropical agriculture and decrease the hazard of disease. All such work required infrastructure – of collections, museums and botanical gardens, and these were established in the colonial capitals as well as in the imperial centres. These enterprises remained central to imperial development through the mid-twentieth century, and to the status of postcolonial powers thereafter (Farley, 1991; Sheets-Pyenson, 1989).

Of course, not all the scientific work of museums and expeditions is to be classed as natural history. We shall see in the next chapter that museum professionals asserted their authority by developing analytical approaches, such as comparative anatomy, and by the end of the nineteenth century some of them pursued experimental methods, for example in the new science of ecology. However, even in universities, and *a fortiori* for city and national collections, the basic rationale remained the documentation and display of diversity, for economic, cultural and political reasons. Though historians of biology are fond of arguing that laboratory science *replaced* natural history in the later nineteenth century (Allen, 1978; Maienschein, 1991), this is an exaggeration. 'Laboratory' biology became hegemonic, as we shall discuss later, but taxonomy remained a vital part of the exploration and exploitation of the world, not least of the European empires that flourished in the first half of the twentieth century.

And so it remains, though the empires are now more obviously commercial. The imperial collections, in all their 'genetic diversity', still represent a major resource for industry, agriculture and 'development studies'. In medicine, too, though most research since the early nineteenth century has been analytical, it required huge amounts of *specimens* and of *data* that were often collected as 'natural history' – for medical 'topographies' of foreign places, as simple statistics on death rates, as accounts of epidemics at home and abroad, or as specimens of human diversity in health and disease. Nowadays, molecular biologists often draw on such material to perform their analyses of variation – in humans and in other organisms – and, as we shall discuss in the final chapter, the ownership of such 'natural resources' can be hotly contested.

But even when natural history was being denigrated by exponents of analysis or experimentation, it remained a major component of *popular* science, an important means by which professional biologists could extend their work and influence, and a key to public understandings of 'nature' – as indeed it remains.

Popular natural history

In surveying eighteenth-century natural history we focused on the landed classes, but such studies were also popular in the 'amateur' middle-class societies that were to be found in many provincial towns by the end of the century, both in Britain and in France. Natural history was at least as important as the physical sciences among the 'scientific' elites of the new industrial towns of Britain (Thackray, 1974). Doctors were often naturalists, and the Industrial Revolution spurred a 'flowering' of field natural history among the industrial bourgeoisie looking for rational entertainment in the countryside. By the 1830s, for example, Manchester boasted a middle-class society (and museum) for natural history, as well as more specialist societies for geology and for botany (including horticulture). In the surrounding towns, later in the century, the study of animals, plants and rocks would usually be combined with archaeology and local history – in the local philosophical society or field club. Though the working classes were allowed into these middle-class organisations on special occasions, for the most part they had their own groups, which often met in public houses. There the local experts would put names to plants; in some cases working men specialised in 'difficult' groups such as grasses or mosses (Allen, 1976; Kargon, 1977; Secord, 1994). For some artisans, botany was connected with herbalism or horticulture, for some it offered social connections, and for many it was a way of exercising talents 'beyond their stations in life'.

Class divisions have softened, but natural history societies (general or specialised) continue to the present. From the late nineteenth century they have sometimes served as volunteer workforces for projects launched and guided by academics interested in (analytical) systematics, or in ecology (an analytical/experimental science of vegetation). Nowadays, with the growth of popular ecological consciousness, local natural history societies may take an interest in pollution, and almost certainly they will be interested in 'conservation'; but still their core activity is the identifying and recording of animals and plants in their locality or region. An enormous literature of guidebooks supports such activities, and museums continue to show shells and stuffed birds – though usually now in ecological stage sets rather than filed in classificatory order like library books. From the origins of photography in the mid-nineteenth century, there has been a strong interaction with natural history, and in our age of conservation, 'slides' substitute for specimens. Films, radio and, especially, television have done much to popularise appreciation of 'nature'; indeed, children in Britain now are more familiar with the ways of dinosaurs (Desmond, 1976) and dolphins than with those of blue tits or badgers.

Displays of technology, new and old

We have already noted that 'spotting' and collecting is not confined to natural objects. Over the last two centuries, as before, certain kinds of machinery have been treasured and recorded – railway engines and aeroplanes especially, but also cars, buses and ships. Though sometimes regarded with suspicion when they were new, such machines came to attract affection, and often became part of the 'landscape', almost part of nature. The industrial age has also seen its new 'collectables': postage stamps are the paradigm case, but cards from cigarette packets were once collected, especially by boys, and sports memorabilia (e.g. football match programmes) now play a similar role. In a world of manufactured commodities and aggressive advertising, human products have become more obvious as objects of pursuit and as avenues for identification, collection and display (Elsner and Cardinal, 1994). I once dragged my younger son around Linnaeus's garden, still preserved in Uppsala, and completely failed to interest him in the principles of botanical taxonomy; he was indifferent to the arrangement of the parts of flowers, but when we passed the cycle shop he displayed an astonishing command of the possible arrangements of the parts of mountain-bikes.

From about 1840, *displays* of technology were a feature of industrial cities; they were intended to instruct the middle classes and/or working classes about the range of machinery (and perhaps about its principles). From mid-century, international displays of technology became common – largely as trade fairs in which countries and companies displayed their latest wares and competed for reputations (and orders). The biological mixed with the mechanical; Richard Owen, the cantankerous zoologist from Lancaster who was to head London's Natural History Museum, was a jury chairman for Raw Materials at the Great Exhibition of 1851, and then at the 1855 exhibition of 'Prepared and Preserved Alimentary Substances' held in Paris! (See Desmond, 1994; *Dictionary of National Biography*, Owen; Rupke, 1994.)

Indeed, the 1851 exhibition is nicely illustrative of the many interconnections of mid-Victorian museums culture. The Museum of Practical Geology was completed that year, near Piccadilly. When the Crystal Palace, built to house the Great Exhibition in Hyde Park, was moved to Sydenham, Owen designed huge model dinosaurs as a public attraction. From the profits of 1851, permanent galleries were established in South Kensington to show the best of British 'design' (now the Victoria and Albert Museum) and of British technology and science (the roots of the London Science Museum) (Pointon, 1994). Behind these promotional displays were the masses of manufacturers' catalogues which are invaluable now if you want to reconstruct the material culture of long-gone industries, trades or professions. Technical libraries, too, were a nineteenth-century creation, including patent libraries (and collections of patent models). From that time, the Western world has supported huge and varied resources for 'extended natural history' – from bureaus of standards to type collections of bacteria or viruses. But, of course, many such 'contemporary' collections became 'historical' with the passage of time – and this tension between the novel and the historical has continued to be a feature of (some) science and industrial museums throughout the twentieth century (Butler, 1992).

Since the Second World War, technology museums have often been established primarily to display *old* technology or industrial archaeology. Many of these museums are on industrial heritage sites – displays of local history and social history. They are not concerned with novelty, nor with the analysis of artefacts or instruments; they show the historical natural history of technology – either by displaying ranges of related artefacts, or by putting artefacts into 'reconstructed settings' (Butler, 1992). In Scandinavia similar displays are to be found in 'folk museums' linked to academic 'folk studies'. In Britain and America, 'folk studies' remained a largely amateur pursuit (part of natural history) (Dorson, 1968).

British collectors, from the mid-nineteenth century, preferred archaeology and anthropology – the former concerned largely with pre-Roman Britain, the latter with the British Empire. Both had their amateurs as well as professionals, and in Victorian Britain much of the activity could be regarded as 'natural history' – collections and classification of artefacts, or customs or languages, etc. A marvellous example of the mixture can still be seen in Oxford, at the back of the Victorian (natural history) Museum (promoted by John Ruskin), where the university houses the Pitt-Rivers Collection of historical and anthropological artefacts. These are displayed by function – e.g. devices for making fire or for scaring spirits – an enormous 'junk shop' arranged like a department store, an assemblage from other cultures which wonderfully echoes the material culture of our own (Chapman, 1985).

'Natural history' now

In this chapter I have tried to trace some histories of a way of knowing I call extended natural history. I have linked description, 'biography,' classification and display and related them both to techniques and to social movements. I have described natural history as a foundation for more complex STM, as an aspect of the economy and as a key mode in which western people have related to their worlds, including worlds that are full of technical devices and concepts. In the final chapter of this book I will return to some of these questions, for our present *c*.2000, but here I would underline two features of natural history which we shall see in the intermediate chapters.

First, the *coexistence* of natural history with other ways of knowing, and the continuing contests around their relative importance – even in museum-based taxonomy, an activity sometimes looked down on as largely 'mechanical'. For example, as we noted, some taxonomists (called 'transformed cladists') now like to assess relationships between specimens by counting common and divergent characteristics; we might say they practise a sophisticated form of Enlightenment natural history. Other classifiers may be more mindful of characters that are regarded as 'fundamental' or 'elementary'; we shall note this possibility when discussing classical (analytical) Victorian botany in Chapter 5. However, most taxonomists since the late nineteenth century have seen classification as a way of exploring evolutionary sequences, an approach which some cladists regard as speculative. Evidently, even in its most technical aspects, natural history can still be controversial. But what of its wider roles and principles?

In our customary, narrow sense of natural history, 'conservation of nature' now seems the ruling principle. Where the early moderns *explored*, and our more recent ancestors *classified*, we

conserve. We worry about loss of habitats in the West and in the
Third World, and about loss of species and genetic variation. For
most of us now, the protection of nature is less a matter of theology than of aesthetics and the precautionary principle
(Bramwell, 1989). Extended natural history, however, is not just
about 'nature', whatever that might mean; it also covers all the
things which men and women create and use, and at this level
the late twentieth century is especially interesting for two
reasons. One is the increasing penetration of global commerce
and the importance of 'brands'; the other is our 'information
systems'.

In this chapter I have stressed the role of commerce – of possession and consumption, and the conceptualisation of craft
objects and of luxury goods as quasi-species. I linked such
'species' to 'natural species' and I contrasted them with 'utilitarian products' (which in the next chapter will be shown to be
products of analysis). But, so pervasive now are 'brands', and so
far are we from the routines of 'traditional' agriculture or any
real exposure to 'nature', that the system of references may be
reversing. The world which is 'natural' for us is not in the fields,
or the natural history museums, or in accounts of the changing
seasons in Hampshire villages, but in our product-packed homes
and gardens, in the supermarkets and the guides to consumption
that flood the television and the magazine racks – not to say the
Internet, by which this commercial development interacts with
the massive increase in the technical capacity for handling 'information'.

Perhaps our World Wide Web can be compared with the
invention of the printed book in the fifteenth century, and of
mass publication in the nineteenth. The former transformed the
distribution of texts and allowed pictorial representation to
become part of the 'descriptive sciences', the latter allowed the
popularisation of natural history and the creation of 'domestic
manuals', etc. At the end of the twentieth century, computer
technology hugely expanded the potential for accumulating and
sorting 'information'; where recording methods can be standardised, specimens or catalogue details can now be compared

worldwide. 'Virtual' collections, virtual museums, mega-libraries all now beckon. The technology is omnivorous, it minimises distinctions; artworks or wild flowers, details of patents or of bacteria, can all be electronically 'digitised' as information. Yet, presently, this world is wonderfully chaotic – you search by index words, not according to a hierarchical classification – *words* are the key. So, magically perhaps, we return to the Renaissance – to a material world that is ordered by 'words', to the power of (brand) names, and to a system of (commercial) enchantment which penetrates our daily havings and doings as once the claims of religion gave meaning to all earthly things. As we suggested in the last chapter, *dis*enchantment is only half the story.

In such ways our present encourages reflexive exploration of the 'material cultures' and 'information systems' of the past. Into such endeavours this preliminary survey of 'extended natural history' may some day fit. To such questions we will return at the end of the book, after we have discussed the characteristic, hegemonic 'ways' of nineteenth-century science – analysis and experimentalism.

Notes

1 For more on monsters see Daston and Park (1998).
2 I thank the author for sending an English version.
3 The phrase is used by Alpers in the title of her chapter 3, and taken from R. Hooke, *Micrographia*, London, 1665, A2^v.
4 For an interesting argument about taxonomies across cultures, see Atran (1990).
5 I owe this point to Marie-Noelle Bourget.

4

Analysis and the rationalisation of production

THINK OF AN EARLY TWENTIETH-CENTURY chemistry laboratory – a big, airy room with rows and rows of bottles full of chemicals. Is it a museum of minerals, of plant and animal products and of the creations of industry? In a way, perhaps, but for the most part, students of chemistry do not enumerate the specimens, worry about the arrangement, look for gaps in the collections or prize new acquisitions, as they might indeed have done in the eighteenth century. Why not? Partly because, since the early nineteenth century, they could buy specimens from companies set up for that purpose, but more fundamentally perhaps, because the chemicals seem to us somewhat 'arbitrary' and lacking in individuality. They are almost all 'compounds', which we can manipulate into other compounds, provided they contain the correct elements. For most purposes each chemical is adequately specified by naming the elements contained and their ratios. We do not want to know how the specimen was obtained – whether from nature or industry. That would only matter if we were worried about the traces of 'impurities' that 'distanced' our sample from being simply what it is supposed to be – a single compound of a few chemical elements. To put it briefly – these chemicals are no longer known through natural history; in the laboratory they are the creatures of chemical analysis.

This chapter and the next trace the formation of analytical sciences, especially around 1800; I stress their relations with professional education and with 'consultancy'. With 'analysis', we approach the knowledges which are esoteric and therefore seen as characteristic of Western science, technology and medicine.

We leave the realms of meaning and those of natural history, and move into expert knowledges that are less continuous with everyday knowledges. In Britain, this kind of science is said to be *not* 'just common sense'. And in the French tradition of history of science associated with Gaston Bachelard, the analytical sciences stand on the far side of the 'epistemological obstacles' which are said to have barred the way to true science, pushing us back into the realms of the everyday with which we are familiar and comfortable. To use one of Bachelard's key examples: the experience of fire was (once) a key part of our phenomenological world; it was freighted with symbolic meaning and literary associations; it was an object of hermeneutics as well as natural history. To talk about fire as the combination of oxygen with other elements, or about the release of specific quantities of energy, was to break from the mundane and enter a world beyond common experience and associations (Bachelard, 1938; Gutting, 1990).

That we can see chemicals in this way is the achievement of the late eighteenth century, though it was not without precedent. This chapter and the next argue that the same historical period created many other analytical sciences, each with their own elements. The elements were not obvious – they needed to be discovered or perhaps invented. I shall argue that Newtonian mechanics, the creation of the seventeenth century, served as a key model; that some new analytical disciplines, notably chemistry, then served as models for others; and that, especially in France, the pioneers of the new developments were self-conscious about method. The key period, *c.*1780–1850, is sometimes described by political historians as the Age of Revolutions – for the French Revolution which began in 1789 and echoed for half a century, and for the Industrial Revolution that spread from Britain. By extending some suggestions in the masterly works of Charles Gillispie, we might also call it the 'age of analysis' (Gillispie, esp. 1965; 1980), and that can be our clue in trying to *explain* the developments.

We shall see that engineers systematically decomposed machines into elementary machines, that other engineers and natural philosophers found how to trace 'mechanical action' through machines, and that engines came to be seen as systems through which the element heat flowed. I shall argue in this chapter that much of what later became physics appeared in the early 1800s at the edges of chemistry and of engineering, as studies of 'elements' such as light, heat, magnetism and voltaic electricity. I include as analysis both the decomposition of 'compounds' into their various elements, and the reduction of systems to the 'flow' of single elements. Analytical chemistry was the model of the first type, thermodynamics could serve as a model of the second. At the end of this chapter, I explore the links between analysis and industrialisation, by considering the 'rationalisation' of production and by musing on the relationships between analytical elements of the new sciences and the 'elements' then apparent in technologies.

In the next chapter I show how animal and human bodies were deconstructed into elements – tissues and then cells – how plants were similarly decomposed, and how these developments were related to changes in medical education and practice. I consider how *form* was analysed in the new 'morphological' sciences, especially in German universities, and how geology became the study of strata (stratigraphy). I also explore the roles of analysis in the new social sciences. But we must not forget the continuing importance of the non-analytical forms, or the *interplays* between analytical knowledges and the 'natural-historical' and/or 'hermeneutic'. Those interplays were central to the politics of science in the age of analysis and they remained so, as I shall try to show.

This chapter and the next focus on the period 1780–1850, when this raft of new analytical sciences and technologies were discovered/constructed. I present them, in large measure, as products of major *institutional* novelties: French professional schools, hospitals and museums, German reformed universities, and the British Industrial Revolution. At the end of Chapter 5, I explore the extent to which institutional and industrial change (and hence wider political and economic factors) may be said to *explain* conceptual change (and vice versa). But I also want to emphasise the importance of later and earlier examples; we continue to create new analytical disciplines – e.g. geonomics. By contrast, some analytical disciplines were constructed in the ancient world and revised in the period we call the scientific revolution. These classical disciplines were about motion – the movements of the planets and, later, the motions of matter on earth. With them I begin.

Analysis from the ancient world

In the chapters on readings and on natural history we mentioned the *meanings* of Greek cosmologies and the Aristotelian tradition of natural history but, in at least one respect, Greek cosmologies drew on other knowledge traditions with very different aims and accomplishments. The Babylonians had recorded the movements of planets and stars in numerical terms – for predictions, not least for astrology. It was Eudoxus, a pupil of Plato, who attempted to draw Babylonian calculations into Greek cosmologies by interpreting numerical ratios in geometrical terms – expressing intervals as circular movements. This mixed tradition was developed by Greek colonists in Egypt. There, in Alexandria, Ptolemy and his school elaborated this line of investigation in the form later transmitted to the Christian West. The movements of the stars and planets could be modelled by a system of about ninety spheres circling round the earth and round each other (Kuhn, 1957).

This way of doing science might usefully be called analytical: skilled, esoteric practitioners (here mathematicians) successfully 'reduced' the baffling motions of the planets to a system of circular movements. The system was not merely descriptive, it predicted motions for calendar-making and astrology. Whether this system was 'real' or whether it was to be regarded as a calculating device to 'save the appearances' became a matter of disputation among medieval and Renaissance scholars; that was part of the discussion around the radical revision of the Ptolemaic scheme suggested by Nicholas Copernicus in 1543, when he showed that a more elegant model could be obtained by putting the sun at the centre of the universe and making the earth into one of its planets.

Between 1543 and about 1700, as noted in Chapter 2, the common Western model of the cosmos was transformed. An Aristotelian model of a closed world gave place to an infinite universe; a complex apparatus of cycles and epicycles was replaced with elliptical orbits around the sun. There was no longer a deep distinction between the celestial and the mundane, and in Newton's great synthesis, celestial motion obeyed the same laws as motion of bodies upon earth. By 1700, in principle, matter in motion obeyed simple laws, wherever and whenever, and this heroic model of analysis set the aspirations for many other realms of knowledge, e.g. chemistry, which were then compounded of natural history and craft, underlain with hypothetical universal explanations in terms of mechanisms or of the Greek elements – earth, air, fire and water.

So early modern 'analytical instruments' might perhaps be seen as models of the world which were also tools for living. Clocks we have met already, as models of the heavens that became regulators of men; perhaps we can add the instruments of the surveyor and the navigator, plus aids to calculation. Perhaps we should see them as the tools of the trades which analysed the earth and the heavens (Bennett, 1986), marking out intervals of time and space. Displayed in an eighteenth-century 'cabinet', they could be said to represent mathematical analysis just as the rows of specimens represented the 'history' of nature and of artefacts, and as the chemists' 'instruments of torture' represented the 'experimental histories' of the Baconian tradition. The practical uses (and prestige value) of the navigational and mathematical instruments ensured a supply and a skill base which could be drawn on by teachers or travelling lecturers who wanted variant instruments or more sophisticated versions (Stewart, 1992).

Analytical aspirations

Those who 'mechanised the world picture' in the seventeenth century separated the primary qualities of matter and motion

from secondary qualities such as colour. In our terms, they separated the 'elements' of a successful analytical programme from
those qualities which seemed less tractable, not least because
they were hard to quantify. They argued that only matter and
motion were sufficiently clear and distinct to be fundamental to
understanding, that here was the deep accord between God and
man; God made clocks of infinite quality.[1] Mechanical models
were speculatively extended to chemistry and to anatomy and
medicine, and they served to stimulate observational studies;
several key microscopists searched the inner structures of plants
and animals hoping to find the 'looms' on which the tissues were
woven – but the results were meagre compared to celestial
mechanics and the programme of 'mechanisation' was failing by
c.1750 (Schofield, 1969).

But as that programme was fading, another was strengthening, which also drew on the successes of celestial mechanics and
especially on the fame of Newton. Instead of emphasising the
mechanical constitution of the world, it emphasised the simplic-
ity of the laws to which apparently complex phenomena could
be reduced. This approach, which developed into the philosophies of science known as positivism, was characteristic of the
late 'Enlightenment'. Ernst Cassirer, in his account of the period,
stressed how the laws of motion served as paradigms of explanation, and how procedures such as the 'resolution of forces'
served as models of *analysis*. The complex motion of a cannon
ball could be resolved into horizontal and vertical components,
and thus expressed as a compound of simple motions (Cassirer,
1955). That model may serve as a simple example of the ambi-
tion to analysis which by 1850 had been successfully generalised
across many new disciplines – not by reducing biology or medicine to matter in motion, but by finding/creating 'elements' by
which new disciplines could be constituted.[2]

Michel Foucault, in his early works, drew attention to a major
fault line in the 'archaeology of knowledge' c.1800; he explored
it for medicine, life sciences, economics and language studies, but
he excluded physical sciences (Foucault, 1970; 1972; Gutting,
1989). Thomas Kuhn, in his essay on mathematical and experi-

mental traditions (Kuhn, 1977b), argued that many new physical sciences took their modern forms at this time; studies of heat, light, electricity and force emerged in mathematical form, alongside the 'reformed classical sciences' such as mechanics. By combining and regrounding the accounts of Foucault and Kuhn, I want to trace the 'displacement' across *most* of STM, including chemistry, engineering and the new physical sciences, as well as medicine, biology, earth sciences and some new social sciences. For many of these new sciences, the new chemistry was exemplary.

The elements of chemistry

We have suggested that eighteenth-century chemistry, like most other Baconian sciences, was a mix of natural history and craft, together with such natural philosophy as might provide speculative explanations of the phenomena. Practical men had continued to advance the chemical arts of medicine, agriculture and manufacturing, as they had done since the Renaissance; savants had tried to effect classifications, modelled on those in botany, or to give explanations of chemical properties based on the Greek elements (earth, air, fire and water) and/or the alchemical principles. But by 1789 the French academician Antoine Lavoisier was offering his *Elements of Chemistry* – a textbook of a new system of chemistry, written in a 'new language'.[3] He drew on new discoveries and new perspectives concerning 'airs'. Common air was no longer a certain quality of air or a blend of qualities, such as the ability to support fire; it was understood as a mixture of various different *kinds* of air, notably those that came to be called oxygen and nitrogen. These could be separately prepared, and had constant qualities, as did other kinds of gas such as hydrogen. Lavoisier allowed these new gases to be elements if they resisted attempts at decomposition. They could be weighed, as could other substances such as metals, which also resisted decomposition. Decomposable substances could then be understood as compounds, and they could be named accordingly.

That this new scheme was worked out in France may be due to the confluence there of the *various* traditions which proved crucial – an interest in the taxonomy of chemical stuffs, information about new gases and reactions, the attraction to quantification and weighing, the widespread concern with 'analytical' understandings and a 'pre-positivist' readiness to define 'elements' in a way which was operational and specific to each new discipline. One can see this methodological and pedagogical programme in the presentation of Lavoisier's work – in a textbook meant to create disciples who would ransack the old chemistry for all they could translate into their new language (Donovan, 1996; Holmes, 1985).

Though Lavoisier and other (salaried) members of the Royal Academy were often called upon to advise the government on technical questions – sometimes at the royal manufactories for munitions, tapestries and porcelain – it seems unlikely that his 'new chemistry' can be seen as deriving substantially from these roles. But after Lavoisier was condemned as a tax farmer and executed on the (new, rational) guillotine, his chemistry was substantially developed by the state's mobilisation of chemists during the wars with the rest of Europe that extended almost without break to 1815. For two decades, France had to be self-sufficient economically and technically while supporting a huge conscript army.[4] We shall see the importance of this conjuncture for engineering and for medicine; for chemistry the effects were perhaps less clear because there were no big teaching institutions dedicated to the subject. For the most part, a group of leading chemists and natural philosophers, some of whom had private laboratories, served as consultants and directors in government workshops and factories. By 1815, they at least were sure that savants in general, and especially the new chemists, had been of huge benefit to the state (Alder, 1997; Crosland, 1967; Fox and Guagnini, 1998: 71–4; Smith, 1979).

The new chemistry led quite quickly to new models of the physical world. The Manchester teacher, John Dalton, soon saw that different elements might have atoms of different weights, and that in compounds (as opposed to mixtures) atoms of different elements might be combined (Cardwell, 1968; Patterson, 1970; Thackray, 1970). The Swedish chemist, J. J. Berzelius, teaching medical chemistry and mineralogy in Stockholm, regularised the new system and introduced the formulae we still know (Melhado, 1981; 1992). The 'operational' potential of this new paradigm was enormous: all the substances in the world waited to be analysed into the new elements, and all the chemical changes in the world awaited description as reactions between elements.

In France and in Sweden, the new chemists were often employed in state institutions for medicine, pharmacy and mineralogy. In Britain, as we shall see, much of the demand came from industry, and chemists acted as 'consultants'. In all these countries, perhaps especially in the USA, agricultural analysis – especially of soils and fertilisers – became a major preoccupation by mid-century. But the key country for the production and reproduction of the new chemistry was Germany, and especially the institute which Justus Liebig developed at the University of Giessen in the 1820s, and which educated so many of the academic and professional analysts of the nineteenth century (Brock, 1997). In that small university town, Liebig successfully institutionalised chemical analysis by drawing on three key factors. He learned the methods in Paris, though the opportunities there for serious students were limited; he applied them in the education of pharmacists – a very important financial support for several key chemists c.1800; and he adapted to laboratory sciences the ideology of collective research cum advanced training which was the pride of the newly founded (or reformed) German universities.

'Research' as the chief aim of universities was then a new idea. As we shall see later, its early practitioners were not physical scientists but philologists who analysed ancient texts, or mathematicians (Olesko, 1988). The German universities reformed or founded after the French occupation, stressed language and *culture* as common features of the many 'German' states, sometimes contrasting them with French analysis. While the French stressed professional education, the Germans stressed broad 'learning' in a community of teachers and scholars – designed to cultivate both knowledge and men (Shaffer, 1990; Turner, 1971). Liebig managed to *combine* the creation of new knowledge with the training of professionals, and by the 1840s he was the acknowledged master of European chemistry. As younger peers and protégés pushed on to new theoretical problems – e.g. how to effect new classifications of compounds, Liebig built up his long-standing links with the world of practice. He published books on the uses of chemistry in agriculture, medicine and industry (Turner, 1982), and his students became consultants in these fields. For the most part their work was analytical – this chemical programme and its progeny probably comprised the nineteenth century's largest investment in laboratory science.

It would not be difficult to show that the age-old interplay between 'chemistry', medicine and the industrial arts had been transformed by the new understandings and practices of analysis. Or that state support for elite chemists in France, for universities in Germany, and for such common national enterprises as mining and the manufacture of gunpowder, all provided resources and incentives for the new chemists. In Britain they depended chiefly on the Scottish universities and the private markets for education and especially for technical advice. By the mid-nineteenth century, British analytical chemists could make a living doing consultancy work in their private laboratories, and maybe teaching part time in the local medical school or 'mechanics' institute' (Russell, *et al.*, 1977; and see Russell, 1996). That the reform of *pharmaceutical* education and practice seems to have been particularly important in several countries (Hufbauer, 1982) parallels the emergence of other forms of analysis in the context of higher professional education. This is clear for engineering, as for medicine, especially in France.

Analysis for engineers

Engineering and mathematics have been related since ancient times. Geometry was a tool for surveyors as well as a clue to the ideal form of the cosmos. If terrestrial mechanics was created in the Renaissance to join celestial mechanics as an analytical science, it was in part because of the interaction between scholars and practical men responsible for fortifications, ballistics, etc. But as in other areas of STM, the reach of analytical methods was limited, and much engineering knowledge continued at the level of natural history and craft recipes.

When, in the mid-eighteenth century, the French government established a school for military engineers, there was no agreed curriculum and no set way of teaching. As Charles Gillispie showed so well, the college was a kind of collective apprenticeship, including mathematics and mechanics. But gradually the teachers took charge of the educational programme, establishing a syllabus of courses which built on previous courses, introducing classroom exercises to supplement lectures, together with examinations to test the student's mastery of each course (rather than just serving for the award of prizes). All these pedagogical devices seem so familiar to us that we need reminding of a world where they had not existed, and of the circumstances of their invention (Gillispie, 1980).

These initiatives in engineering education were pushed by young lecturers keen to make careers and reputations – to produce good engineers, but also to produce scientific work for their own prestige. Textbooks would serve their own students, but they would also be used by other engineers, inside and outside training institutions; they could serve as scientific treaties, but also for 'learning at a distance' (Picon, 1992; 1996; Weiss, 1982). It was in this context that Gaspard Monge and others developed a series of methods that would be central to engineering education down to the present. They extended Renaissance insights into analytical disciplines, including the 'projective geometry' which later generations would learn as 'technical drawing'. Such drawings became crucial to engineering practice, but they also had another significance for teaching institutions – they helped bring a wider range of machines and constructions into college. For much the same reasons, such colleges, like medical schools, developed *collections* of machines;

museums became a part of this practical professional education. For big machines and civil constructions such as bridges, scale models were used – to allow vicarious experience without travel, to facilitate comparisons, and to help make clear the key features of the machine in question.

But if machines were collected and compared, what was to be the framework of comparison? Should it be *function*, comparing different kinds of prime mover, say? Such classifications had been used in some contemporary anatomical museums (e.g. that of John Hunter in London) and they were to remain a staple of technological museums. But could you dissect machines (as you could bodies) and find their elements – the 'simple machines' from which all others were made? An interest in this question was clear in Renaissance mechanics: Galileo wrote about the wheel, the lever, the screw, etc. as simple constituents of more complex machines; but in engineering, as in medicine, it was in the decades around 1800 that the first systematic texts appeared, breaking down machines into elementary parts and presenting the resultant classifications of machines as an authoritative form of engineering understanding.

The authors of the text regarded as seminal were Jean-Marie Lanz, a young Frenchman, and Agustin de Betancourt, a Spaniard who had supervised the construction of hundreds of models intended for a new technical college in Madrid. I do not know the extent to which Betancourt drew on other contemporary STM, or on the contemporary vogue for analysis as a general method, but such comparisons were explicit for one of the founders of advanced technical education in German states (Perez and Tascon, 1991). Von Reuleaux, in the mid-nineteenth century, drew parallels between his analyses of machines and comparative anatomy; he drew a second parallel with chemistry, and was so led to devise a nomenclature for 'elements of machines' by analogy with the agreed formulas for chemical elements. As in chemistry, one function of this nomenclature was to devise *possible* compounds that might be *synthesised* (Reuleaux, 1876).

By then, analysis of machines (and projective geometry) had

become mainstays of German technical education. They were particularly useful for pedagogues who wished to assert that higher technical education was practical but also had the rigour expected of traditional *university* disciplines. In some such colleges, machine elements were available in standardised form, to be assembled by students who thereby developed the 'analytical eye'.

Analysis and 'physics'

Historians of physics have tended to see nineteenth-century physics as an extension of natural philosophy and mathematics rather than of the new chemistry or engineering, but perspectives are changing. It would be worth checking how many of the nineteenth-century experimental physicists had trained as chemists.[5] Donald Cardwell and his Manchester associates showed that 'energy' studies in Britain owed much to engineers (Cardwell, 1971; 1989; Pacey, 1974), and Crosbie Smith and Norton Wise have continued the lesson more recently by investigating the career of William Thomson, Lord Kelvin, both in Glasgow University and among the industrialists and marine engineers of that Second City of Empire (Smith and Wise, 1989). And it was graduates of the Ecole Polytechnique, the school for military engineers established in 1793 after the Revolution, who *made* French physics *c.*1830 (Fox, 1992; Geison, 1984; Gillispie, 1980; Shinn, 1980). Perhaps we should take the engineering tradition more seriously – not just for the analysis of machines into components, but for the analysis of 'flows' of 'elements' through machines and through 'nature'. The elements here were 'mechanical action', 'force', 'heat' or, later, 'energy'; the central machine was the steam engine.[6]

We know little about the invention of the first practical steam engine, devised by Thomas Newcomen to pump water from Cornish tin mines. Later eighteenth-century improvements, especially the separate condenser, were made by James Watt, an instrument-maker in Glasgow University who worked for James

Black and knew his studies on heat. Black, by elucidating 'latent heat' and in separating 'quantity of heat' from temperature, had greatly advanced the analysis of heat and this was of advantage to Watt (Basalla, 1988; Cardwell, 1971; Donovan, 1975; Hills, 1989; Jacob, 1997).

It was Lazare Carnot, one of the founders of the Ecole Polytechnique, who provided the foundational analysis of 'mechanical action' in machines. The question at issue was how 'force' could be transmitted most efficiently through a set of cogs and levers (say). Carnot showed that the transmission was most effective when it was smooth, without jarring of the parts; in his *Essai sur les machines en générale*, he focused on the 'work' done by machines (Gillispie, 1971). At about the same time, the British engineers who were emerging as key players in the Industrial Revolution stressed 'mechanical power' or 'duty' as a measure of what the new steam engines could do. Thereafter we can trace the two 'national traditions' to the energy doctrines of the mid-nineteenth century. In Britain, practical engineers and their philosophical associates reflected on the potential of engines of various kinds; formal teaching of engineers was scarcely an issue until after mid-century (Buchanan, 1989; Cardwell, 1957). In France a series of investigators, schooled in mathematics and the analytical techniques of the Ecole Polytechnique, created new disciplines expressed in textbooks and in papers to the Academy of Sciences in Paris (which was *not* a major centre of industry).

In France, the generation after Lazare Carnot, including his nephew Sadi Carnot, created mathematical analyses of heat flow in machines. They used the mathematics developed by Joseph Fourier, a graduate of the Ecole Polytechnique who had rejected the 'microphysical' models by which Laplace and his colleagues in the Academy of Sciences tried to extend Newtonian analysis into a general theory of attraction between all particles of matter. By contrast, Fourier's 'theory of heat' was essentially macro-scopic, geometrical and practical – the engineer's view from *outside* the machine, an assessment of what a machine could *do* (Grattan-Guiness, 1981; Grattan-Guiness and Ravetz, 1972; Smith, 1990: 328).

By contrast, in early nineteenth-century England, there was little professional education in engineering (or in chemistry). The technical and philosophical analyses took place chiefly in the 'societies' of Scotland, London or Manchester, where practitioners with a taste for analysis, and teachers oriented to practice, interacted informally and through publications. Engineers who made and operated steam engines concerned themselves with their 'duty', or horse power – their capacity to drive pumps or machinery and therefore their economic efficiency (Morrell and Thackray, 1981; Smith, 1998). In industrial Manchester, where 'duty' was well known, we can trace these concerns through to the 1840s and the work of James Prescott Joule who extended considerations of efficiency to the new electric motors, and to whom we owe the principle of the mechanical equivalent of heat.

By trade, Joule was a brewer, by inclination a cautious investigator; politically, he was a Conservative in a Liberal-radical town. He had been raised to science as a pupil of John Dalton, who had achieved world fame for interpreting French chemical elements in terms of atoms. By contact with Dalton's friends, and at meetings of the Manchester Literary and Philosophical Society, Joule became familiar with problems of engine efficiency and with the tradition of 'modelling' machines to investigate such issues. His experiments which became classic were provoked by popular enthusiasm for 'electric motors', touted in the 1830s as capable of indefinite speed and power output. Joule was sceptical. He set up model electric motors to measure the 'work' produced by a given amount of zinc in the battery. (All this technology was *very* new.) He concluded that electric power was, in fact, rather expensive compared to the benefits of burning coal in a steam engine. He went on to measure the relationship of mechanical action to heat production, using a 'paddle wheel' immersed in water heated by its rotation. In all these experiments Joule was measuring relationships, assuming that 'force' or 'action' could not be destroyed, but only converted into some other form (Cardwell, 1989).

The two traditions came together in Britain when reformers of 'natural philosophy' imported French methods. At Cambridge, the capital of British mathematics, French methods were central to a revival movement c.1830, led by young graduates interested in engineering and industry. Charles Babbage, later a 'philosopher of industry' and the inventor of a proto-computer, was prominent, but it was in Glasgow that the new mathematics chiefly met the new engineering. The only British industrial city which then had a university, created a professorship in engineering in 1840, and six years later they appointed to the chair of natural philosophy, the young mathematician and experimenter, William Thomson, later Lord Kelvin. William's father was a teacher of mathematics, his brother James was a pupil of the Glasgow engineering professor and then an engineer in a London shipyard and marine engineering works. William had studied mathematics at Cambridge and spent some time working with Victor Regnault in Paris, a chemist turned physical experimenter who taught advanced measurement techniques and conducted extensive researches on the relation between pressure, temperature and specific heats; his data was simplified into 'tables' for engineers – a characteristic (analytical) guide to practice. Appointed to Glasgow at the age of twenty-two, Thomson also focused on the analysis of steam engines, and on analogous analyses of electric motors. The work expended or absorbed in an electrical engine depended on the potential difference, just as it depended on the temperature difference between the boiler and condenser of a steam engine, or the height of fall for a waterwheel (Smith, 1990; Smith and Wise, 1989).

This theory about the flow of heat was based on French analyses to which Joule had objected on metaphysical grounds: they seemed to allow the 'annihilation of force' which for Joule was a power belonging to God alone. Joule had proposed that the 'force' was converted into heat. Thomson, by 1851 was able to reconcile those views by the principle of 'conservation of energy'. Though it is well known that other scientists arrived at this principle by other means (Kuhn, 1977a), I emphasise here that the best recent accounts of the key French and British traditions demonstrate the plausibility of seeing 'energy physics' as based in engineering analyses; the principles of nature could be derived from machines. But how were manufacturing-machines made and understood; what kinds of analysis helped *produce* them? In what ways were the factories of the Industrial Revolution a matter of analysis?

Rationalised production

We return here to the points I made in the introductory chapter to this book, about the relation between ways of knowing and ways of making – especially to the relation between analysis and the rationalisation of manufacturing production. Perhaps we can see the Industrial Revolution as extending the practice of analysis from the divisions of time and space in the early modern period to the divisions of labour in the modern age. Arguably, a mechanical loom was to weaving as a clock had been to the heavenly motions – a materialisation of analysis.

Here we can follow Edmund Cartwright, the clergyman and natural philosopher who invented the mechanical loom in the early years of the Industrial Revolution. For him an automatic loom was not just an enticing economic proposition, it was a philosophical challenge: only if he could mechanise weaving, could he be sure he understood it. By analysing the process into simple human actions, he could substitute mechanical actions that together would reconstitute the process (O'Brien, Griffiths and Hunt, 1996). Thus understood, looms belong to the world not just of clocks but of the automata which had fascinated eighteenth-century high society – those doubly enchanting products of a deconstructive, disenchanting analysis.

But, importantly, 'rationalised production' was not necessarily 'mechanisation' because not all 'machines' were made of metal or wood; they could be made of *people*, treated as mere components. Indeed, it could be argued that the mechanisation of people was historically prior to the construction of inanimate machines. Lewis Mumford, in a seminal article on 'Authoritarian and Democratic Technics' (Mumford, 1964), long ago suggested that the sets of slaves who built the pyramids should be seen as precursors of mechanised production. Adam Smith's classical analysis of increased productivity in eighteenth-century Britain referred not to new machines but to 'division of labour' in workshops; in pin-making, to quote the famous example in his *The Wealth Of Nations* (1776), the separation of tasks between different workers allowed each task to be performed more efficiently. Here, indeed, was the basis for mechanisation; but in one sense the key step had already been taken – the elements of the productive process had been separated, each had been developed as the sole work of one component, and these components had been assembled to form a productive mechanism which reflected and expanded the work once carried out by a single skilled craftsman (Babbage, 1832; Berg, 1979; 1980; Marx, 1967).

For such reasons, we should use 'rationalised production' (rather than mechanisation), as a basic 'type' of 'making', allowing that the elements were usually then mechanised, more or less. This rationalisation of production, from the industrial revolution factories to the production lines of Henry Ford and the 'ergonomics' and 'scientific management' of Gilbreth and F. W. Taylor, lies at the core of 'industrialisation' over the last two centuries .[7] (Though here, too, we should remember that old and new crafts were also involved – e.g. crafts of machine-making or of canny maintenance (Braverman, 1974).) Relentless analysis and the standardisation of machines and of men have been central to most dystopic visions of modern manufacture and of modern society more generally.

Rationalisation and identities

But how are we to relate the 'scientific' analysis with which we began this chapter and the 'technological' analysis now under discussion? If Cartwright analysed the action of a weaver, or the English farmer, Robert Bakewell, the desirable properties of a sheep carcass,[8] they provided frames for systematic discussion and improvement, but they did not thereby establish new 'sciences', as did Lavoisier with his new chemical elements, or Xavier Bichat (as we shall see) with his tissues of the body.

How their contemporaries saw that difference, and how we feel it, deserves more attention than I can devote to the task. Perhaps realist philosophers have already shown how it is that 'elements' became 'real' rather than merely instrumental. The problem is doubled in as much as some elements that seemed philosophically secure, e.g. 'heat' or 'caloric', were later turned into mere 'properties' of more substantial elements – heat was not a 'stuff', but rather the motion of other kinds of molecules. From the end of the nineteenth century, even the familiar chemical elements lost their specificity. In principle, and under peculiar conditions, one could find several kinds of phosphorus, say; each chemical 'element' had become a collection of 'isotopes', a range of possibilities in a matrix of subatomic particles.

This shift from a 'kind' to a 'range' may seem remote from industry but, in fact, it was also a key aspect of rationalisation of production. Here my initial guide was an article which explored understandings of textiles in France from the eighteenth century through the early nineteenth century. In a nice instantiation, corresponding to my general argument about craft-objects and natural-historical species, William Reddy showed how the system of local guild production of textiles required specifications of local cloths. These were printed in books that read like natural histories, but were produced by an inspectorate. Though with the rise of unregulated, extra-guild production these accounts lost relevance, they continued to be produced *faute de mieux* (Reddy, 1986). Only after the onset of industrialisation was a new genre of text printed; and these were about production rather than specific products, about mechanised spinning and weaving, about the capacities and versatility of machines. Of course, information about products continued to be available in France as in Britain, but in the form of catalogues and sample books (the obvious equivalents of floras, or collections of pressed plants). In a sense, the specific products had become points (or groupings) on the grid of possibilities which machines could realise.

We may note the similar contrast between reproducing pedigree animals and the more utilitarian forms of breeding which
were designed to secure new sets of characteristics.[9] In Britain
from 1750, cattle and sheep were changed radically as breeders
sought marketable characteristics and faster growth. Sheep,
Bakewell said, were machines for turning grass into money;
they could be redesigned to do so more efficiently. And that new
mutability of domestic animals was well known both to Charles
Darwin and to his contemporary, Gregor Mendel, the
Moravian monk and science teacher who in the 1860s discovered mathematical ratios in the inheritance of plant characteristics. Indeed, both the theory of evolution by natural selection
and Mendel's work on inheritance can be shown to have built
on this shift in breeding technology (Desmond and Moore,
1992; Orel, 1984).

In all these cases, 'species' – whether manufactured, chemical
or biological – lost something of their 'given-ness', and came to
appear as points (or clusters) on a matrix. In my terms, they
moved from natural history to analysis and rationalised production. But note that pedigree animals, like wines and luxury
brands, remained as quasi-species.

Production and analytical sciences

From this discussion of production and products there emerges
a picture which complements our account of analytical ways of
knowing. We may see 'technological' analyses of productive
processes, whether mechanical, biological or human, running
alongside 'philosophical' analyses of 'model' situations.
'Technological' here refers to direct analysis of technologies –
e.g. by testing waterwheels, measuring by-products of chemical
plants, or keeping data on rates of fattening in cattle. In all those
cases, practical investigators looked for regularities that could
guide understandings and improvements. Statistical trials of
drugs in hospitals fall into the same class if conducted on the
basis of 'suck it and see' (Matthews, 1995).

More 'philosophical' investigators sought to uncover stable 'elements' behind natural and artificial phenomena – usually looking for them in special circumstances, for example in laboratories where phenomena could be simplified, dissected and measured more easily. The elements included those of the chemists, the tissues of the body (and eventually, Mendel's units of inheritance, which became genes). Such elements could then be used for elucidating productive processes – though often it was difficult to link analysis-in-principle with operations-in-practice, and many industrial processes continued to be operated according to 'technical' rules of thumb – i.e. by natural history and craft.

If, then, this analysis of analysis is persuasive, we may need to rewrite the history of physical sciences so as to give more weight to worldly practices, especially engineering. The historical relation of physics to engineering may indeed come to resemble that between physiology and medicine, or of chemistry and heavy industry. In all cases the 'science' provided a language in which to analyse healthy or unhealthy 'workings'. These languages of analysis were crucial to 'consultants' and to teachers who made a living out of expounding and exemplifying the principles, especially in the institutions of higher professional education in France and Germany. In these 'places of learning', analysis took precedence over craft (Layton, 1971; 1974).

Such analyses were also important to practitioners, especially when there were 'difficulties'. But though consultant analysts provided diagnoses and suggestions, they did not often 'produce new practice'. In the 'places of practice', e.g. factories and English teaching hospitals, practitioners were usually in charge on the basis of 'experience'; analysis, there, was for keen junior staff or consultants called in on special occasions.

In private industry and especially in Britain, graduate chemists worked chiefly as consultants. By analysing the inputs, outputs and by-products of industrial processes they could effect large savings, as Andrew Carnegie was to discover in his steelworks (Mowery and Rosenberg, 1989: 28–34). Many analytical chemists were self-employed, typically they were 'liberal' scientific professionals (Russell, 1977; 1996). But by the end of the nineteenth century they were also working for government in the enforcement of industrial and public health legislation that they had helped promote. The first inspector of 'pollution' under the British Alkali Act was Robert Angus Smith, a Manchester chemist now best known for discovering 'acid rain'.[10] Chemists also featured as experts in patent disputes. In all these roles – improving production, assessing pollution, devising regulations, etc. – chemists functioned chiefly as *analysts*; and for the most part that is what was taught in institutions of chemical education, including the courses on technological chemistry which began to appear in technical colleges, especially in Germany.

Mechanical and civil engineering also offered opportunities for consultancy – for assessing new bridges or dams, or for standardising boilers (Buchanan, 1989). Some of the work was for the firms concerned, some for government, some for insurance companies. To take one key example, the devastating effects of boiler explosions prompted much expert analysis of the safety requirements, and insurance regulations played a major role in establishing reliable safety standards. Such roles brought mechanical engineers into the business of legislation. Civil engineers had long been involved, advising on schemes for roads, canals, railways and reservoirs, which usually needed parliamentary permission.

We shall deal in later chapters with the growth of 'electrotechnics', but can note here that the massive expansion of electrical industries at the end of the century was based in part on new techniques of electrical analysis and standardisation, on new instruments robust enough to work outside laboratories and on technicians able to diagnose and repair electrical circuits and machinery. Those were the skills taught in the new technical colleges and in the university departments of physics and electrical engineering which mushroomed to meet the demand and/or to benefit from the excitement (Fox and Guagnini, 1999: ch. 3). (Much the same could probably be said of the electronic industry several decades later.) By seeing most chemists and consultant engineers as analysts, we clarify their roles, their social relations and their common features, and we prepare ourselves to recognise the huge importance of these activities down to our present.

We shall return to analysis and industry in Chapter 7, but we turn now to analysis in medicine and the life sciences, and in the sciences of the earth and of society. Again we backtrack to the seventeenth century, before focusing on the relevant professionals in the decades around 1800 – after which we follow the patterns through to the twentieth century.

Notes

1 For first-rate recent accounts of the scientific revolution, see Henry (1997), Schuster (1990) and Shapin (1996).
2 I am indebted to the seminal but unpublished PhD by Randall Albury, which showed in 1972 the relations between the elements of chemistry and of medical analysis. See also Albury (1977).
3 For an accessible and authorative history of chemistry, and for full further references, see Brock (1992).
4 For a survey, see Dhombres (1989).
5 Faraday, Magnus, Regnault and G. C. Foster (of University College, London) might start the list; and Helmholtz moved to physics from physiology. See also Nye (1996).
6 See the very useful article by Crosbie Smith (1990).
7 On rationalised production in France and the US military see Alder (1997) and Smith (1985). On scientific management Cohen (1997).
8 For agricultural rationalisation see Thompson (1978).
9 I am indebted here to the work of Isabel Phillips (1989) on styles of animal breeding in the nineteenth century.
10 For more on Smith and his fellow chemists, see Farrar (1997) and Gibson and Farrar (1974).

5

The elements of bodies, earth and society

GO ROUND A HOSPITAL LABORATORY and count the ways to examine the elements of bodies – by post-mortem anatomies or microscopical specimens, by chemical analysis of blood and solid tissues, by immunological testing for antibodies, by electrical sensors such as electrocardiograms (ECGs) and electroencephalograms (EEGs), or by imaging techniques from X-rays to magnetic-resonance scanners. For 200 years we have been finding new ways to analyse bodies (Amsterdamska and Hiddinga, 2000).[1]

This chapter explores the genealogies of these forms of analysis, and the related analyses of plant and animal structures, processes and forms. It also includes earth sciences, and the social sciences which were invented c.1800. It complements the previous chapter and together they map the massive restructurings of STM which were effected from the late eighteenth century. At the end of this chapter I explore ways in which we might attain some *explanation* of these creations and their consequences.

Medical analysis: corpse and patient

Major public dissections were highlights of academic medicine from the Renaissance. Initially, they were meant to illustrate Greek understandings of how the body worked; they exemplified the classical analytical traditions associated with Aristotle and Galen – the humours which made up the body, the central role of the liver in converting food to venous blood, and of the left side of the heart in transmitting the 'breath of life' from the lungs to the arterial blood. They became explorations of the bodily landscapes, a kind of natural history which revealed a landscape with many new features, including the lymphatic vessels full of watery fluid – the rivers of the body. But anatomies were also associated with a radical revision of part of the classical analysis. Just as in planetary astronomy, where the sun replaced the earth at the centre of the world, so by the demonstrations of the English physician William Harvey, the relations of the body organs were turned upside down. In Harvey's book of 1628, *On the Motion of the Heart and of the Blood*, the heart became central, pumping blood round and round the body, where previously blood had been supposed to seep gently from its source in the liver into the flesh to which it gave substance. But what was the circulation for? The lack of adequate answer was one reason for restricting the enquiry to '*matters of fact*' rather than the 'purposes' of the Greek tradition (French, 1994; Pomata, 1996; and see Cunningham, 1997).

As in the physical sciences, this radical revision of the classical analysis prompted much investigation and, though the discovery grew from an Aristotelian tradition, the restructured organism was often presented as mechanical; René Descartes wrote an essay on the body as a mechanism, centred on the heart as a pump. But though the mechanical analogy proved fruitful in analysing the operations of the muscles and bones (Cook, 1990), the effects on medicine generally were minor. Some of the explanations changed, but the practice of medicine remained much the same; its form remained biographical and natural historical.

By the mid-eighteenth century, in physiology as in chemistry and microscopy, the attempt to analyse bodies as matter in motion was running out of steam. Instead, investigators began to examine the general properties of living matter, sometimes as they appeared in the 'simplest of living forms' such as the polyp or *Hydra*, which served the eighteenth century much as the *Amoeba* served the nineteenth and early twentieth (Canguilhem, 1975; Delaporte, 1994; Jacob, 1988; Roger, 1971; Schofield,

1969). As in the physical sciences, investigators began to look for
the elements of *particular domains*, rather than assuming that
the first principles of *all nature* were mechanical, or the old
Greek elements. In the creation of new biomedical, earth and
social sciences, as for those discussed in the last chapter, the aspirations of the technical professions were crucial.

By the later eighteenth century, anatomy had become the
science of surgeons. Elite surgeons, individually or collectively,
were investing their wealth in the 'cultural capital' of anatomical museums, often associated with the dissecting rooms where
their students practised anatomy 'hands on'. In most European
countries, leading surgeons were approaching the status of
physicians – not just by anatomising, but by using public hospitals for surgical operations, by proving useful to military authorities, and generally by appealing as 'practical men' to the
expanding bourgeoisie. As their influence grew, especially in hospital practice, they popularised their more anatomical understandings of disease (Fissell, 1991; Lawrence, 1996). However,
because surgeons usually remained subordinate to physicians,
their anatomical approach still operated within the hegemony of
a natural-historical, biographical medicine based on systemic
disturbance, whether of humours, fibres or nerves. More generally, the control exercised by lay-governors over charity hospitals in Britain, and of religious orders over the public hospitals
of France, ensured that pastoral care usually took precedence
over the acquisition of knowledge (Risse, 1999).

It was in France, as a result of the Revolution, that the power-
orderings were reversed, at least in a few key teaching institutions. Political change was rapid and deep, and so was
institutional change in medicine (and science and technology).
Old power structures, including the Paris Faculty of Medicine,
were swept away, and with them the support for classical medicine of a literary and biographical form. Aspirations were
released – for a medicine which *unified* physic and surgery; for a
more practical, more surgical, medicine, taught by the best qualified to the most promising; and for medical control over the hospital patients, alive and dead.

In Paris, before the Revolution, the poorhouses were run by Catholic nuns; in the Hôtel Dieu, just by the cathedral of Notre Dame, thousands of sick poor were tended unto death. Some of Paris's leading surgeons were attached to the hospital, and they tried to use the inmates for the systematic teaching of their students and apprentices – but the nuns resisted the intrusion. After the Revolution, however, the power of the Church was reduced and doctors gained control over the running of the hospitals. They were able to arrange the patients according to their diseases, conduct classes on the wards, use students as assistants and conduct autopsies on the dead paupers. It was in and around these 'medicalised' poorhouses that analytical medicine was invented and tissues became the new elements of the body (Gelfand, 1980; Risse, 1999; Vess, 1974).

The inventors were teachers associated with the new School of Health (later the School of Medicine), established in 1794 by a Revolutionary government bombarded with complaints about quackery and in desperate need of doctors for a conscript army now at war with the rest of Europe. Leading surgeons, physicians and medical reformers were recruited to staff the new school and to devise a curriculum. Initially, they improvised to meet the emergency; then they worked out a programme which promoted anatomical notions of disease as the basis for the common training of both physicians and surgeons. Students and staff were selected by competitive examination and supported by the State. The opportunity was enormously exciting – a chance to lead the reform of medicine. Paris became the centre of the world for medical students and investigators (and retained that status for fifty years). In the School of Medicine, as for engineering in the post-Revolutionary military polytechnic, dreams of analysis now coincided with the resources to realise them (Gelfand, 1980; Duffin, 1998; Maulitz, 1987; Risse, 1999; Vess, 1974).

The hospitals supplied lots of bodies, and young surgeons dissected hundreds of them. The medical students frequented the Museum of Natural History and learned comparative anatomy; they knew about the recent triumphs of French chemists, especially Lavoisier; they also read philosophy, then called *ideology* and aiming to provide a scientific account of how ideas were formed and employed. As a science of man, ideology was some-times regarded as a branch of medicine, and it taught the impor-tance of analysis as a way of knowing (Albury, 1972; 1977).

We know that the brightest of the young surgeons, Xavier Bichat, intended his 'tissues' as the elements of general anatomy. They were the basic structures, homogeneous to the eye, from which the human body was composed; usually twenty-two were listed, including bone, skeletal muscle, liver, etc. The idea was not altogether new: Aristotle's works included discussions of het-erogeneous and homogeneous parts, and some British anatomists had focused on membranes as common to many dif-ferent organs, but Bichat generalised the questions and provided systematic answers. He catalogued the tissues, exploring their 'sensibility' and their 'contractility', and their reactions to chem-ical reagents. He also studied the ways they became diseased (Haigh, 1984; Pickstone, 1981).

It was in *pathological* anatomy that tissues proved most fruit-ful for medicine, providing a framework for the view of *diseases as lesions*. Diseased corpses could be systematically surveyed to see which tissues were affected, and how. The new geography of the body proved profitable because lesions in an organ were often restricted to particular tissues, e.g. pericarditis – inflam-mation of the membrane around the heart. Or inflammation might affect similar tissues in various organs – sometimes because of the continuity of tissue systems, e.g. of the lymphatic vessels or the veins (Maulitz, 1987). Throughout the nineteenth century, the analysis of tissue lesions was a major part of medi-cine; in big city hospitals 'pathologists' came to be employed full time to dissect and to teach. From the 1830s microscopes were deployed, and soon thereafter, as we shall see below, it became possible to analyse tissues and lesions in terms of their con-stituent *cells*. We have already seen that the new chemistry could be applied to medicine, for the analysis of normal and diseased fluids and tissues. So now we begin to see the genealogies of the many analytical methods which I noted in the introduction to this chapter. But the corpse-analysis of the Paris teaching hospitals had a further extension – to the clinical examination of the living patient.

When diseases were but disturbances of equilibrium, searching for them in the body was no more rational than searching for the site of hyperinflation in a present-day economy. But if diseases were lesions – inflammation of tissues, say – they could in principle be *found*. Maybe they could be located and assessed before death; the patient could be analysed before it became a corpse. And sure enough, the Paris hospitals which demonstrated tissue analysis to their students also took up auscultation ('sounding' the chest), and then the *stethoscope* – a wooden tube between the body of the patient and the ear of the doctor. By transmitted sounds, a doctor could 'see' the diseased lungs or heart; he was trained to correlate sounds with the pathology revealed at autopsy. Thus clinical examination began to take its familiar form. In nineteenth-century teaching hospitals diagnosis was all; therapeutics was a sideshow. Doctors in the big continental hospitals were often sceptical about customary remedies, inclined to let diseases 'run their courses' – the patterns of pathology were clearer that way.

The ramifications of analysis for medical organisation proved very extensive. In as much as diseases were specific, anatomical and *local*, doctors could legitimately specialise in particular organ systems. That was the basis of many of the nineteenth-century special hospitals – for diseases of the eye, etc. – often resisted by general physicians and general surgeons, but pushed by ambitious young men anxious to corner a part of the medical market. In Britain such medical entrepreneurs often boasted of training in Paris, where medical teaching was so diversified as to offer lots of arenas for the development and dissemination of specialist knowledges (e.g. Granshaw, 1985). And if diseases were specific and defined, you might treat them 'en bloc' in hospital wards and compare the numerical results of different treatments. The 'numerical method' of the Paris hospitals, though not entirely without precedent, is the basis of hospital statistics as developed in the twentieth century to provide measures of efficacy in medicine. Statistical analysis was built on pathological analysis, in hospitals (and also in 'public health') (Matthews, 1995).

The medical alternatives to analysis

So far, I have told a unidirectional tale of the advance of analysis in medicine, as if the analytical medicine of (some) teaching hospitals was the whole story. Indeed, history of medicine is often presented in this way – 'hospital medicine' replaced 'bedside medicine' (and was itself replaced by laboratory medicine).[2] But this is a misleading travesty. Thinking about ways of knowing, and ways of making, allows a more complex, political and helpful story, for medicine and for other technologies.

As I have tried to stress in earlier chapters, natural-historical or biographical medicine was not *replaced*, even in teaching hospitals. For decades after the Revolution, many French hospitals continued to practise a medicine that was chiefly biographical. In Britain, the charity hospitals in which medical teaching was conducted remained wary of appearing as places of dissection and 'experiment', for fear of dissection after death was widespread in Britain c.1830, and the 'resurrection' of corpses was a huge scandal, even before Burke and Hare were discovered 'making corpses' for Edinburgh anatomists (Richardson, 1988). In any case, British medicine was dominated throughout the nineteenth century by physicians and surgeons whose chief pride was their lucrative private practices. They did not see themselves as medical scientists; they taught bedside skills that would prove useful when their protégés set up as general practitioners (and referred their difficult, rich, cases to their old masters). The system did not produce pathological discoveries at the same rate as the big continental teaching hospitals, but it may well have provided better doctors and better care for the urban poor.

In Victorian Britain, 'medical science' tended to be the preserve of young men, not yet established as consultants, eager to win a name as teachers, or to impress professional societies by showing their findings or by importing continental knowledge. They aspired to be 'analysts' in a world of 'experience' (of craft and natural history). By the end of the century, a few such would-be medical scientists had gained salaried posts as anatomists, physiologists or pathologists in medical schools. They included the pioneers of bacteriology – a new form of medical analysis which was taken up by public health doctors (a characteristic linkage of technical analysis and social management).[3] These 'scientists' in medicine fought to make medical schools more like science faculties, including full-time clinical professorships; and they dominated the Medical Research Council (MRC) founded by the government just before the First World War. They then used the MRC to push standardisation (and experimentalism) in medicine, but not until the mid-twentieth century would their way of working become dominant among elite hospital clinicians. Throughout their long campaign for a more analytical and experimental medicine, they drew on continental, and then American, examples.

In Austria and Germany, some of the big-city teaching hospitals (Vienna and then Berlin) had been strongholds of analysis from the early nineteenth century (Duffin, 1998), but most medical schools were in universities in small country towns. Had they wanted to adopt 'mass' pathological anatomy, they did not have the numbers of patients required for efficiency. In fact, many early nineteenth-century German physicians saw a future, not in mass dissections but in the ever closer examination of individual patients. In a culture that placed great stress on 'development' – in embryos, lives or societies – charting the unfolding of diseases in the biographies of patients was both intellectually fulfilling and a practical way of teaching. In the later nineteenth century, with the rise of experimental physiology, laboratory methods could be bent to these ends. Physiological medicine could be more individual, less 'museological', than the pathology of Paris and Vienna (Tuchman, 1993).

Doctors in the USA drew on all these European traditions. In the early nineteenth century, many Americans had studied in Paris, especially for anatomy and surgery; later medical students went mostly to German universities, to learn pathology, physiology and bacteriology. By the end of the century, the returnees had managed to create a few full-time posts for medical scientists, even for scientific clinicians. Backed by big philanthropists such as Rockefeller and Carnegie, whose agents were much impressed by science, they created German-type departments and hospitals in a few East Coast cities. The pattern spread rapidly in the twentieth century, partly to raise the status of doctors by restricting access to medical schools; and Americans promoted it in Britain from about 1910 (Warner, 1991; 1994).

However, in all countries, in as much as doctors dealt with individual patients and helped guide their lives, biographical medicine necessarily remained part of practice. Even in the most soulless of continental teaching hospitals, case histories were taken, if only to check the results of the physical examination, and where patients were paying for treatment, no doctor could afford to be too analytical. Young American doctors were warned *not* to bring home the doctor–patient relations they saw in the Paris poorhouses; new knowledge gained from hospital autopsies could be useful, but it had to be subordinated to the consideration of the individual, paying, patient (Warner, 1985).

We see that the relations between analytical and natural-historical medicine were always complex, dynamic and context-dependent (as they were also in other kinds of STM). In the twentieth century, as analytical methods developed further, some doctors reacted by advocating a return to bedside medicine, or to a more 'holistic' approach or to the study of 'the natural history of diseases'. These movements were significant in all the major countries between the wars, and since the Second World War we have seen recurrent attempts to stress 'biographical' aspects of general practice, including psychosomatic and psychosocial dimensions.

The dialectic is inevitable and healthy (Lawrence and Weisz, 1998); in all practical activities there will be tension between the claims of formal analysis and those of 'experience' which are hard to formulate; we see such tensions in industry and agriculture as well as in medicine. One advantage of 'ways of knowing' is that such problems can be seen 'across fields', comparatively, so that we can learn from a wider range of other cases.

In the next section we return to Paris *c.* 1800, to examine analysis in the biological sciences, but bear in mind that these were often closely related to medicine, from their origins to our present.

Analysing plants and animals

The National Museum of Natural History had been the Royal Botanical Garden (and Menageries). Because natural history was well regarded by the Revolutionary government, and because those responsible for the Royal Gardens were politically nimble, the garden was not abolished with the other royal institutions of the *ancien régime*. Instead it was 'nationalised' and its staff expanded by the appointment of new specialists, who were meant to form a republic of professors, each with their own section of nature (Spary, 1995). That was not the outcome. By 1800 Georges Cuvier dominated the institution, at least on its animal side; he pioneered a structural-functionalist analysis of animals, though his colleagues championed very different approaches – as we shall see.

Cuvier's job, as a curator-professor and a rising member of the scientific associations of Paris, was to discipline his domain – to order it, but also to show that studying animals was as scientific as the new chemistry, or the mathematical physics then championed by Pierre Simon Laplace. For this purpose, zoology could no longer be a matter of arranging snail shells – Cuvier would leave this to his colleague Lamarck, who annoyed him by claiming that animal species could change into other animal species. To Cuvier such 'transformism' was nonsense, based on a superficial natural history that failed to get to grips with comparative anatomy and the deep-seated, structural differences between animals. He came to argue that the animal kingdom comprised four chief groups, each with a defined basic plan – radiates (jellyfish, sea urchins, etc.), molluscs, segmented invertebrates (e.g. worms and insects) and vertebrate animals. To understand how animals were alive was to understand the relations of structural and functional relations between organs, *and*, additionally and crucially, the functional interactions between each animal system and its *milieu*. (This is the origin of our meaning of 'environment' as more that just 'surroundings'.)

For Cuvier and his followers, the study of animals was no longer surface-taxonomy plus extrapolations of human physiology. In the new zoology, as in chemistry, the combinations and properties of elements served to explain both structure *and* process, variety *and* its classification (Coleman, 1964; Outram, 1984). Here was the basis of a new 'professional' science for museum curators and university professors. As we shall see below, it was developed most fully in the German universities, and from the later nineteenth century in those of Britain and the USA; it was the mainstay of the zoology curriculum in high schools up to the 1960s. The same kind of analysis was applied to plants, so constituting 'classical botany', a discipline whose trajectory paralleled that of zoology.

In 1800, plant classification had been more developed than that of animals. The comparative anatomy of plants was also better known, in part because 'their anatomy was external' – the main parts (root, stem, leaves, flower parts, etc.) were easily seen in the adult plant and even in the seed. Parisian botanists stressed the different *plans* of plants, notably the deep differences between monocotyledonous plants (such as grasses and palms) and dicotyledonous plants (such as buttercups or oak trees) – the former grew from the base and the leaves had parallel veins, the latter grew from the tips and the leaf veins branched. This division was not to be read as simply an accumulation of differences, but rather as a matter of fundamentally different '*organisation*'. It was akin to Cuvier's focus on structure and function, but the emphasis was on structure – partly because plant physiology was obscure. It was in Paris that a Swiss botanist, Auguste-Pyrame de Candolle, worked out a classification which has remained as the basis of plant taxonomy. It was harder to use than that of Linnaeus, but seemed more natural; professionals liked it, amateurs differed, because they preferred a practical scheme over analytical niceties (Daniels, 1967). In some ways de Candolle and the German professors of botany were followers of Cuvier, but they also drew on 'morphology', that analysis of form which Cuvier had come to reject (Daudin, 1926; Morton, 1981; Sachs, 1890).

Analysing form

In principle there are at least two ways of analysing structures. So far in this chapter we have discussed the multiple-element-compound model – chemicals, medical bodies, animals and plants are made up of *various* elements, tissues or organs, compounded in different ways with different outcomes. That was the French mode of analysis – it called for 'dissection'. But there was another mode, often followed by those for whom dissection was 'murder' – a mode seen as German and as opposed to the analytical spirit of Paris. It involved the perception of a structural *idea* which 'unfolded' through development; the complex organism, in its whole and its parts, expressed this idea. The best known advocate of this view was Goethe, the German polymath and poet, who saw leaves as the fundamental organs of plants. (The parts of flowers were modified leaves – so, too, were stems and roots.) A plant was not primarily a compound of elements but a transformation and multiplication of a simple element (Cunningham and Jardine, 1990; Lenoir, 1982; Russell, 1916).

That this position and Cuvier's were not entirely opposed may be evident to us, but antagonism grew as morphologists in Paris, notably Cuvier's colleague Etienne Geoffroy Saint-Hilaire, argued for fundamental similarity not just within each of Cuvier's four main kinds of animals but across them. Geoffroy saw arthropods as animals with skeletons on the outside, as vertebrates inside out; Cuvier was appalled and there was a famous showdown at the Paris Academy in 1830, a year of political revolution in France. Cuvier won the debate 'on points', but died soon afterwards; Geoffroy, as champion of the Romantic left enjoyed a burst of popularity. In France, as in Britain, morphology appealed to political radicals, who often linked it with Lamarck's version of transformism to provide a materialist explanation of the variety and 'adaptedness' of plants and animals – an alternative to the natural theology of the clerics.

Morphology has been overshadowed by 'evolution' but it was then a highly productive science, asking questions about form scarcely thought of before. For example, morphologists were concerned with symmetries: why did the front and back limbs of vertebrates have the same basic structure? Was the skull a series of vertebrae, much modified? These questions remained fertile for much of the nineteenth century, they were often linked to questions about development and embryology, and later they became part of 'evolutionary theory'. But they were initially and logically quite separate from evolution as a historical process. Indeed, much of what is *now* understood as evidence for evolution was first discovered as evidence for the morphologists' 'common vertebrate plan'. As an example we can cite the homologies between ear-ossicles and gill arches, i.e. the demonstrable continuity of form and position across all the intermediaries between the tiny ear-bones of mammals and the bones which support the gills in fish (Appel, 1987).

The study of animal form was also pursued at the microscopic level, and was linked with the medical microscopy mentioned above. The 'general anatomy' of Bichat used tissues as elements, but many anatomists, especially in the German universities began to look for common 'generative' tissues from which the other tissues could be derived. Here again was the Romantic emphasis on unity and common ancestry; here, too, was an incentive to microscopy. Could one find a microstructure common to different tissues? Candidates for that role included the idea that all tissues were made of uniform globules of protein, but this view collapsed when better microscopes were developed in the 1830s (Pickstone, 1973). A new candidate emerged in 1839 from the joint researches of a botanist, Mathias Schleiden, and a young anatomist/physiologist, Theodor Schwann. They argued not so much for a common *unit*, as for a common mode of development: that in all plants and animals, including man, the fluids of the body 'crystallised' to form 'nuclei' around which proteinaceous materials were deposited. Plant cells acquired walls and fluid vacuoles appeared in the cell-stuff; animal cells did not get walls, and they adopted a wide variety of shapes – from flat skin-cells to elongated muscle fibres. Here was another huge analytical programme, called histology, to excite anyone who could master the use of a microscope: all the tissues of all the plants and animals in the world – adult or embryonic, healthy or diseased – could be analysed in terms of the development of cells (Ackerknecht, 1953; Bynum, 1994; Harris, 1998).

Readers who know modern biology may have spotted a catch. *We* do not believe that cells are formed out of fluids, any more than we believe that microbes can be so formed; for us cells come from cells, and microbes from microbes. In so knowing we follow the French and German investigators who revised cell theory, not least by examining simple plants, and who argued that microbes were tiny plants produced by other tiny plants. And if we combine these two new perspectives, we see one of the great generalisations of the mid-nineteenth century (and a new meaning of eggs) – all animals and plants were made up of cells, derived from other cells and ultimately from an egg-cell. That egg-cell came from the mother and in sexual reproduction had been fertilised by a sperm-cell (or pollen-cell) from the 'father'. Now follow back the ancestry – individuals came from parents, cells from cells, as far back as the species went. That grand analytical conception is so familiar to us that we forget it is only 150 years old.

We have already suggested that morphology in general, including embryology and histology, drew much of its strength from the concern of German culture with *development*. Through cell-theory it became closely entwined with studies of plant and animal organisation, and with questions of evolution, as we will briefly explore in the next section.

Classical zoology and botany

What did zoologists and botanists *do* in the nineteenth century? From many histories, including much of the best recent history of science, one would guess that they collected and classified as

part of natural history, that a few of them pioneered experimentalism in biology. and that their overwhelming concern was with
evolution, especially debates about Darwinism. This is a distortion, arising in large measure from the 'peculiarities of the
English' and the Anglo-American orientation of much of the historical writing. Before about 1870, Britain had very few professional biologists, but it was a world leader for imperial
exploration, amateur natural history and natural theology – thus
it produced Darwin. The impact of the *Origin of Species*, when
published in 1859, was enormous; it dominated later nineteenth-
century discussions of natural theology and the relations of
science and morals. In our day, Darwin's disciples have made
evolutionary theory and sociobiology into lively sciences and
major components of 'public understanding' of science. But as
Peter Bowler has shown repeatedly, writing against the tide of the
'Darwin Industry', if you read what professional biologists were
publishing between 1860 and 1940, Darwinism is much less
prominent than one might suppose (Bowler, e.g. 1983; 1988).[4] I
think we can illuminate the situation by remembering that most
of the biological sciences were then not just natural history, nor
experimentalist; they were analytical. (I have already outlined
some of them, and more will follow.)

Consider the problems that Darwin faced as a naturalist who
also knew the work of French and German zoologists. Some
were geographical – how could you analyse the patterns of
animal and plant distribution in terms of climate etc, and in
terms of the territories of similar species (Browne, 1983;
Nicolson, 1987)? Some were morphological – why were the
basic structures of large groups of animals so similar when their
modes of life were so different? Why were embryos of different
kinds of animals more alike than were the adults? Some were
taxonomic – why did some groups of animals and plants seem
much more varied than others, at the level of major groupings
and at the level of subspecies? Could species of animals and
plants be represented as 'budding off' from previous species?
Such questions were *not* matters of simple observation. They
arose from the biologists who had analysed animal forms, shown

how embryos diverged, analysed patterns of classification, and tried to work out laws for the distribution of animal species in space and time. We may see Darwin as pulling together all these forms of analysis and showing how their results could be explained if, indeed, animal species were derived from each other over long periods of time. He was by no means the first to think of that, but he made the case very systematically, using all the best science of his time. And he provided a possible mechanism, natural selection, which many professionals thought insufficient but which most recognised as serious possibility (see Bowler, 1989a; Browne, 1995; Desmond and Moore, 1992).

In this perspective, evolution was to the analytical sciences then sometimes called biology, as energy was to the analytical sciences sometimes called physics. The new doctrines were ways of pulling together previously separate domains, of finding a deeper common ground. In a way that was new for biology, if not for the more mathematical sciences, they were achievements of 'theory', co-ordinating and reorienting existing analytical fields. In an 'age of capital', as the political historian Eric Hobsbawm has suggested, these doctrines were impressive 'accumulations', built on the creations of the 'age of revolution' (Hobsbawm, 1962; 1975). Indeed, the doctrine of evolution, when combined with the 'revised' cell-theory discussed above, meant that all animals and plants could be seen as parts of an enormous lineage of cells, now extending back in time beyond the origin of the species concerned, back to the common ancestor, back to the origin or origins of life. The study of that great genealogy was 'biology', a potential intellectual synthesis of many disciplines held together by these new generalisations.

But just as physicists went on analysing the behaviour of light and heat, electricity, magnetism, etc., so zoologists and botanists went on analysing form and development, usually now at the microscopic level. They went on trying to relate embryology to taxonomy and to palaeontology, as they had done since the early century. They used their museums for comparative studies, and the laboratories for dissections and microscopical studies of structure and development (Bowler, 1996). They worked, for the

most part, as analysts – and some of them, as we shall see in the
next chapter, also came to incorporate the methods of experimentalism.

It is easy now to overlook the continuing significance of analytical STM, partly because of the success of 'experimentalists'
from the later nineteenth century. When 'experiment' was proclaimed as essential to science, analysis could seem old-fash-
ioned, as we shall see in Chapter 6. The analysts rarely asserted
themselves against the experimentalists' simple contrast of
experiment with 'mere description'; so it came to be taken for
granted. Nowadays, if asked about modern molecular biology,
for example, most participants and commentators, even specialist historians, would probably describe an *experimental* science,
in contrast to the old kind of biology which was natural history
– mere description and classification. But that opposition is
naive; to my mind it leaves little room for most of the science of
the nineteenth century and much of that which followed.
Physiological chemistry, say, or classical embryology are much
more than 'mere description'; the double-helix structure of DNA
was revealed, not by experiment as such, but by the systematic
mobilisation of a range of *analytical techniques* – notably
organic analysis of the bases and X-ray crystallography of suitable crystals. Watson and Crick were successful because they
realised the importance of the problem, and because they were
well placed to collect the relevant analytical results and to see
how they could be fitted together (Abir-Am, 1997; Olby, 1974;
Watson, 1968); they did few, if any, 'experiments'.

Sciences of the earth

Eighteenth-century 'geology' had comprised special crafts plus
parts of natural history and natural philosophy. Minerals, soils,
etc. were described and classified, and stories were offered about
earth histories. The former enterprise was continuous with the
natural history of plants, the latter with general theories about
the nature of the cosmos. There was no separate discipline of

'geology'. That discipline was created as teachers of mining and mining surveyors came to envisage the rocky mantle of the earth as a series of different layers or strata. These were not all to be found at any given site, and sometimes they appeared in the 'wrong order', but a right order could be discovered – an ideal series of strata in terms of which the strata of any given site could be understood. That hypothesis proved productive, strata proved to be reasonably distinct – they did not generally grade into each other, they could be recognised by their mineral compositions (and later, especially, by their *fossils*).

Stratigraphy was a boom science of the early nineteenth century, giving intellectual interest to the appreciation of landscape and promising to be useful for mining industries. Hunters of fossils could classify them by an extension of Cuvier's zoology, or use them to characterise strata (and then to wonder about the succession of life forms). Thus stratigraphy led back into new questions about earth history, and thence into the discussions of evolution which featured so largely in later nineteenth-century biology (and in most historical writings on geology and biology). But as a form of scientific work, as subsidised by governments to facilitate mining, geology was stratigraphical puzzle-solving. Imperial interests extended that mapping project worldwide (Laudan, 1987; 1990; Rudwick, 1972; 1976; 1985; Secord, 1986).

Mineralogy was another new earth science; it came in two kinds according to the elements used. Chemical mineralogy dissected minerals in terms of the chemical elements; crystallography was a kind of morphology, based on the discovery of 'unit crystals' – solid geometrical forms characteristic of particular minerals. Again, the Paris Museum of Natural History was a key site (Laudan, 1987; Metzger, 1918), and again there were questions about the power of professor-curators to take the specimens apart – some of the mineral specimens were 'jewels'.

But stratigraphy and mineralogy were not the only forms of earth analysis that became part of exploration and imperialism from the early nineteenth century. The patterns of the earth's magnetic fields were mapped, and so were isotherms – lines of equal temperature; all of which enterprises involved the collection of data from many points on the globe. The chief instigator of several such projects was the cosmopolitan German, Alexander von Humboldt (whose brother Wilhelm was one of the chief authors of German university reform). Alexander was well connected in all the scientific capitals of Europe and he travelled widely in America. He was one of the people who mapped patterns of vegetation, in relation to climate, taking account of altitude as well as latitude. It was Humboldt who called for systematic observation of geomagnetism, and Britain was one of the countries to respond. Under the control of the British army, observation stations were established in Montreal, Tasmania, the Cape of Good Hope and Bombay (Bowler, 1992; Browne, 1983; Cannon, 1978; Nicolson, 1987).

That the British navy was heavily engaged in mapping coastal waters throughout the world, is part of the background to the Charles Darwin story. HMS *Beagle*, on which he voyaged in the early 1830s, was a naval ship; the captain, Fitzroy, was an authority on barometric measurements, and there was a naturalist paid to take the trip. (Darwin was the unofficial naturalist, helping reduce the social isolation of the captain for the years away from civilisation.)

As the English natural philosopher John Herschel noted *c.*1830, all these extensions of the world of knowledge added to the richness and to the speed of scientific development which seemed part of that analytical age (Herschel, 1835: ch. 6). And the speed and richness increased further as steamships and telegraphy improved communications. Analytical meteorology became much easier, and more useful, when widespread data on weather, especially barometric pressures, could be collected simultaneously. With such data, the elements of the science as we know it were worked out – the cyclones and the anticyclones that we see each day on the weather forecasts.

In the twentieth century, 'aerology' – the study of the atmosphere as a complex dynamic system was pioneered by a Norwegian physicist equipped with sophisticated mathematics (Bowler, 1992). At about the same time, the Manchester physicist Patrick Blackett was collecting data on the geomagnetism of rocks in the ocean floor, so lending support to the once heretical theory of continental drift, and helping initiate the 'tectonic revolution' in earth sciences (Hallam, 1973). He was also encouraging Bernard Lovell to develop cosmic ray studies by constructing a huge radio telescope at Jodrell Bank (Agar, 1998; Lovell, 1976). All these projects were primarily analytical, rather than experimental, indeed they contained much that might be called natural history; they involved observatories and field measurements more than laboratories. They illustrate my general argument that new analytical forms of science have continued to develop and will continue to do so – in academic settings, in the field and 'at work'. But now, again, we return to 1800, to explore the origins of the social sciences.

Analysing the social

One of Michel Foucault's strongest claims about human knowledge was that the human sciences were invented about 1800, and might now be on the way out (Foucault, 1970: ch. 10). I would argue that new social sciences were indeed created, in analytical mode, alongside the new natural sciences. They too were created 'out of' natural-historical knowledges, which were displaced rather than replaced; the natural history continued to matter, even when the new analytical forms were hegemonic.

Foucault showed that eighteenth-century discourses about wealth and language, like those about animals and plants, were mostly of the kind I call natural history; they were classificatory, focused on systems of correspondences, easily expressed in *tableaux*. By contrast, the new sciences of political economy and philology were about 'organisms' – economies or languages with characteristic structures and patterns of change. But were these new sciences analytic in using 'elements', and might the new patterns of work have anything in common with those we have discussed for natural sciences?

Whereas eighteenth-century discourses on wealth were mostly

descriptive or prescriptive, the political economy established in
Britain *c*.1820 by the author and financier, David Ricardo, was
analytical in its reduction to such key notions as agricultural
rents and the productivity of land, or the relations of supply,
demand and mechanisation. It is no accident that here we first
find the simple diagrams of idealised relations that continue to
decorate economics' texts and which look so much like the diagrams in texts of physical chemistry. Analytical economics,
henceforth, would be substantially deductive; 'realistic' accounts
of business development or national economies have continued
to be closer to natural history than to analytic formalisms. This
tension is recurrent; one sees it in the late nineteenth-century disputes between 'English' analytic accounts and the sociological
studies of the German school; or in British universities now,
where prestige economics is mostly mathematical, and less
abstract forms are mostly classed as 'development' or 'management science' (Roll, 1973).

Whether historical sociology can help explain the emergence
of analytical economics is an open question. In as much as it was
first a product of the Scottish universities, it shared a pedagogical context with the emergent medical and chemical sciences. But
the chief characteristic of early nineteenth-century political
economy in Britain would seem to be its very high public profile
and politicisation; Ricardo's formulae, and the related demography of T. R. Malthus, were issues of the day. Ricardo's works on
agriculture, subsistence, and the problems of pauperism, like his
work on 'the machinery question', were central to political discussion. He taught a nation the principles by which new problems might be managed to the satisfaction of the propertied
classes. For this, the authority of 'science' mattered; it was bolstered by the formalism, by the mobilisation of statistics, and by
this double similarity to the new physical sciences with which it
was closely associated.

Several well-known authors wrote on chemistry or geology, as
well as on political economy; the same publishers, journals and
institutions 'carried' all these sciences; all taught respect for the
'laws of nature and society', both to the bourgeoisie who bene-

fited thereby and to the working classes who did not (Berg, 1980). As Foucault stressed, the political economy of the early nineteenth century was doubly related to the emergent sciences of life: through the demography of Malthus that focused on human reproduction, and through the questions of agricultural productivity that were central to Ricardian economics (Young, 1985). French 'sociology', as invented by Saint-Simon and Comte was closely related to 'biology' through models of organisation (Haines, 1978; Pickstone, 1981). That the emergent science of philology was also so associated may further the general argument for simultaneous reshaping across the range of STM.

The study of language had been focused on relations between words; it came to explore the differing structures of languages and the rules of their historical transformations: 'the isolation of the Indo-European languages, the constitution of a comparative grammar, the study of inflection, the formulation of the laws of vowel gradation and consonantal changes – in short the whole body of philological work accomplished by Grimm, Schlegel, Rask and Bopp' (Foucault, 1970: 281). Comparative grammar was explicitly analogous to comparative anatomy; and the study of language groups was part of the new study of human variation that also encompassed physical anthropology and aspects of the new archaeology.

There appears to be little work on the social and material aspects of this birth of philology, perhaps, because we easily assume that such science can be done anywhere. But if we examine the best English history of philology, one guided by Kuhnian paradigms rather than by Foucault or materialist sociology, then we notice, more than did the author, the significance of libraries, and especially of the rare collections of exotic manuscripts, e.g. in Sanskrit or Norse, that were investigated by the new philologists (Aarsleff, 1983). Previous writers on language had relied largely on the skills and literary knowledge familiar to most scholars, but the new philology was based on collections that could be investigated in much the same way as collections of pressed plants or animal skeletons. Characteristically, in the new or reformed universities of Germany, institutes created for the study of philology had *research libraries* and *seminars* for the discussion of research results – two of the tools of scholarship to which we are now almost blind – two keys, perhaps, to the wide extension of analysis across the disciplines. As other historians have pointed out quite independently, the tradition of 'research' which was to make German universities famous, began in philology and in mathematics, rather than in the physical sciences (Turner, 1971).

There is room for much extension of the argument. For example, we might regard mathematical statistics as the new form of analysis of numerical data, using means and standard deviations, etc. as ways of conveying the essential characteristics of the aggregate (Hacking, 1990). Demographers, in turn, found statistical ways of characterising populations and their development. And some of the nineteenth-century's new 'subjects' could be regarded as analytical 'compounds'. For example, the new study of public health was primarily about characterising places – not by natural history or topography, but by drawing on analytical chemistry and stratigraphy as well as on mortality statistics (Hamlin, 1998; Pickstone, 1992b). Phrenology, the science of reading human propensities from the shape of the skull, was the new analytical science of character, based on the location of particular faculties in particular parts of the brain (Cooter, 1984). Nineteenth-century periodicals were full of analytical discourses which we could see as running in parallel and overlapping – e.g. political economy, public health and geology (Rudwick, 1980). Which is not to say that natural history, topography, physiognomy, case histories or simple sets of numbers were excluded; these natural historical forms continued, but in new relations. As we have seen for analytical political economy versus the historical school, the relationships were contested.

By the end of the nineteenth century, even in Britain, 'social sciences' were established in key universities. The neo-classical, marginalist economics of W. S. Jevons, Alfred Marshall, etc. was developed in universities – in Manchester, in Cambridge, and from the early twentieth century in the London School of Economics (LSE). We can treat neo-classicism, with its emphasis on markets as the means of determining prices, as a new form of analysis. Like its classical parent, it was related to contemporary natural sciences – here to kinetic theories in physics (Mirowski, 1989; Roll, 1973; Schabas, 1990). But the decades around 1900 also saw the development of higher education for business managers; this was much less analytical and mathematical – it comprised economic history and the methods of accountancy; after the First World War it could include the physiology of fatigue, much researched among the workers in munitions factories.

We see similar tensions in several countries since the Second World War as government funding for a huge expansion of universities and healthy demand for 'non-science' graduates, allowed substantial expansion of the social sciences. For the first quarter-century post-war, much of the emphasis was on the analytical aspects of social sciences – on kinship and structural-functionalism in anthropology, on social structure and mobility in sociology, and on formal linguistics and formalist accounts of literature, etc. The last quarter of the century saw two opposite pulls. Scholars in university departments developed more hermeneutic approaches in anthropology, sociology and cultural studies – the so-called linguistic turn, but among the growing ranks of 'management scientists', there was a renewal of more historical, descriptive and indeed prescriptive studies that we might call natural history – loosely linked with such weakly analytical forms of social sciences as occupational psychology and 'theories of the firm'.

But, of course, 'natural historical' and hermeneutic accounts of social phenomena were not, and are not, the prerogative of academics or indeed of 'investigators'. In social investigations, the 'objects' can speak. They have their own systems of meaning, their own 'natural histories' and, in some cases, their own modes of analysis. I will touch on these issues in the final chapter, but first I want to pull together some of the issues about 'analysis' raised in this chapter and the last, so also preparing for the chapter on experimentalism.

Reflections on the institutions of analysis

Readers who have attended to the various passing remarks about the institutional sites of analysis will have noted that the sections on the decades around 1800 referred frequently to higher professional education and to state museums, hospitals, etc., especially in France, and that later sections referred repeatedly to universities, especially in Germany. From time to time I referred to observatories, surveys, expeditions and field stations. I also stressed the analytical work of 'consultants' in industry, agriculture and medicine, the place of analysis and rationalised production in factories, and the roles of analysts in industrial research laboratories. What causal claims, if any are here involved? How am I picturing the relations between these various 'institutions' and the ideal-type way of knowing that I call analytical?

Perhaps I need first to say that no account of such complex questions can be both simple and adequate, and that I draw from the best studies of such matters the general lesson that all such causation is reciprocal. People's ideas obviously influence the institutions they create and the ways the institutions develop but, equally, the institutional patterns in which men and women are raised and work also structure the ways in which they think and act, at a conscious level but also through lots of assumptions and tacit knowledge that the people concerned may never examine. This interplay means that we can reasonably speak of ways of knowing (and of making) as *embodied* in the workings of institutions, where 'institution' can cover surveys or factories or, indeed, the working habits of a group of investigators, provided they are sufficiently ingrained and stable to be passed on to any new members of the group. So, for example, if we consider the STM institutions of major European countries *c*. 1850, I would want to claim that the major teaching hospitals, national museums, state surveys, professional schools and 'advanced factories' can sensibly be characterised as chiefly analytical. Staff in such institutions worked on people, specimens, natural features, processes, machines or economies – which they diagnosed, ordered and regulated according to their elements.

Do I want to claim that the formal institutions concerned – e.g. the Museum of Natural History in Paris, or London teaching hospitals or engineering workshops in Manchester necessarily *originated* as analytical institutions? No. It is a historical fact that some of them originated for other reasons and had different 'modes of operation'. Indeed, in the chapter on natural history I argued that the big state museums grew as 'treasure houses' and places of display more than for systematic study, and that where collections were used for systematic study and education, this was mostly as natural history. The same is true of teaching hospitals; in the later eighteenth century 'walking the wards' became a prized part of medical education, especially in London, but the cases were 'displayed' as individuals; their 'massing' in hospitals was a matter of convenience – you could see many individual patients in a short time. We have noted that in some early engineering colleges, education was by a sort of collective apprenticeship, and the layout of many engineering workshops had no obvious rationale in terms of the steps of production. Analysis then loomed small in these collections of specimens, patients, tools and pupils.

But such collections could be (re)made to facilitate analytical enterprises. They could be reordered, materially and/or conceptually, to promote and display analytical knowledges and practices, provided that people who were so directed and motivated had the conceptual and material resources and the social power to effect these transformations and implement new ways of working. By sketching the long history of analysis as an ideal, I have suggested one root for the aspiration. By stressing the discovery of 'elements' constitutive of the new fields of knowing, I underlined that aspiration alone will not suffice; creative processes were required that were to some extent unpredictable. But what of the motivations and the power?

For industrial and agricultural analysis the motive seems relatively straightforward – the expectation of material gain by improving productivity through better control of inputs, outputs, processes and by-products. Such expectations may not always have been realised; many commercial processes remained very resistant to analysis – in which case, craft and lore ('natural history') were better guides. But the expectations were well met in many cases, enough it seems to gradually convince industrialists that analysis was worth paying for, at least when things went wrong. Analysts, of course, actively encouraged such convictions.

For governments, too, economic gain or increased efficiency more generally could be major motives. British politicians of the early nineteenth century were mostly educated in classics rather than natural philosophy, and so were the generals and the admirals, but they recognised a national (and class) interest in the improvement of navigation, agriculture or mining and were persuaded to support data collections and analysis which promised such improvements. When faced with escalating costs and/or unrest over such matters as the relief of the poor and the containment of epidemics, they were willing to lend political clout to the intellectual authority claimed by the economists and their 'reformist' doctor friends. In such ways old institutions were partially converted to analysis, and new institutions, such as the geological survey or the Poor Law Commission, were designed to the plans of analysts (Hamlin, 1998).

In the French professional schools and the German technical schools (as, later in the science faculties of British and American universities), new analytical knowledges were often produced to elucidate the 'principles of practice'. By the end of the century, all major universities and technical colleges had laboratories where chemical apparatus, gas-engines or dynamos were provided so that students could learn the techniques they would later apply 'on site', in 'works' laboratories' or in private 'testing' facilities. All such professional education was closely linked to worlds of practice and, by 1900, to government institutions for regulation and standardisation of practice, such as the National Physical Laboratory (1900) and its German model, the Physikalische-Technische Reichanstalt (Cahan, 1989; Magnello, 2000; Moseley, 1978). This was the institutional network that produced and used analysis. (We shall return to it in Chapter 7; perhaps we could think of it as the analytical version of technoscience.)

Yet I would stress again that natural historical knowledges continued to be important. We have detailed the argument for medicine, but it also applies to other parts of STM. The debates between medical analysts and 'biographical medicine' were paralleled in those between industrial analysts and upholders of craft knowledges In taxonomy, we have mentioned the tensions between professional botanists much taken with de Candolle's structural classification and those who preferred the easy Linnaen system (Daniels, 1967). Such disputes were common in the nineteenth century, and they remain important. So also for contested boundaries between different kinds of analysts. How, for example, should geological chemistry be related to chemical geology, or how did X-rays compare with tapping and the stethoscope as a way of 'looking into the chest'? Whenever a new analytical technique is invented, the 'classification' which results needs to be squared with existing classifications, and professional 'turf' may be at issue. Moreover, since many analytical techniques came to be practised by technicians rather than the 'principal investigator' or the doctor in charge of the case – so additional boundaries came to be contested.

In stressing analysis as a way of knowing I am trying to highlight features which are common to many fields of STM and over many periods; thus we can facilitate comparative use of data and insights across different cases. Although my chapters trace particular 'ways of knowing' over time, I emphasise again that these are *all* to be understood as *always* developing in dynamic, often contested, relationships – *both between and within the ideal types.*

In the next chapter we turn to experimentalism and invention. These seem to me to have a social dynamic different from analysis: they were not, generally speaking, so closely related to the dissection and regulation of economic and professional practices. Rather, I shall argue, they were developed in 'sanctuaries' deliberately distanced from professional and commercial practices, privileged places where novelties could be created and controlled. These sites were equipped as for analysis, but were associated with incentives to synthesis – rather as early analytical sites had been equipped for natural history but converted to analysis.

Again, it is not hard to see why, in the ever more complex world of nineteenth-century industry, some inventors might try to set up invention-factories, or why some industrialists might relieve their chemists of routine analytical work so as to concentrate on creating new products. But what of the practice of making experiments – how and why were sanctuaries provided that were both academic and 'constructive'? Part of the answer, especially for France and Britain in the early nineteenth century, lay in a few peculiar institutions such as the Royal Institution and the Collège de France, where professors, often trained in analysis, lectured to fashionable audiences attracted by demonstrations of power over nature. However, the chief answer lay in universities, especially in Germany. By mid-century, partly because of their successes as analysts, many German professors of science had well-equipped laboratories; they also had the time to do as their discipline seemed to require. They did not have to spend all their days on teaching or on analysis of specimens, they could try to build knowledge, so expanding the 'research ethic' of the universities. Relying on the training of students for analysis, but not needing to be at the beck and call of practitioners or industrialists, they could create model controlled-worlds, so attracting students and other academics who could be persuaded that such syntheses were a step beyond mere analyses, and that laboratory novelties were more attractive than the messy world outside.

Notes

1 For very useful histories of pathological analysis, see Foster (1961; 1983).

2 The important works by Jewson (1974; 1976) are often read in this way.

3 On germs, see Worboys (2000); on science and management, Sturdy and Cooter (1998); on transformations of British medicine, Lawrence (1994).

4 For a discriminating survey, see Bowler (1989a).

6

Experimentalism and invention

CONSIDER LOUIS PASTEUR'S swan-necked flasks, deployed in public experiments in mid-nineteenth-century Paris. The drawn-out necks were bent so that any air passing into the flasks would deposit its 'dust' or organic particles on the bend of the neck; it would not reach the broth in the bowl of the flask. The design was a clever solution to a problem in fermentation experiments – how might you allow the access of air without the access of dust? By this means Pasteur showed that 'clean' air did not cause fermentation but dust did. By many such experiments he satisfied his audiences that he could 'control' fermentation. It was *not* a spontaneous process; microbes were not spontaneously generated (at least under everyday conditions); microbes entered from the air and they caused fermentation in organic broths. Of course, you could not prove that microbe-generation was never, ever spontaneous, but you could marginalise questions about the ultimate origin of microbes (or diseases); you could move them beyond the normal concerns of chemists, brewers, hygienists and doctors.[1]

This control over fermentation served to explain some inventions and to provide many new ones. It made clear why, if you boiled broths and jellies in full vessels and then capped them to exclude air, the fluids would usually 'keep' until the jar was opened. This was the principle of 'canning', which proved so important for the importation of meat from South America. And the heating routines which Pasteur worked out were used to sterilise, or pasteurise, milk or wine with minimal change to the taste.

Much of Pasteur's work was rooted in industrial problems; his work on fermentation derived in part from alliances with beer-makers in Lille, and his later work on diseases from epidemics which killed silkworms. Arguably, the force of Pasteur's research programmes derived from his training in chemical experimentalism which he applied to 'model diseases' of wine and beer. Fermentations, he argued, were biological rather than chemical, but the techniques were not far removed from his chemical studies. Then he pursued his experimentalism into diseases of animals, when human diseases, for obvious reasons, were not the subjects of direct experiment. Pasteur exploited 'biotechnologies' which could be manipulated – his 'models' gave rise to practices as well as principles. That work continued in the Pasteur Institutes erected to his honour in Paris and the French colonies; their studies of bacteria and vaccines were 'state of the art' in biomedicine around 1900.

This chapter argues that experimental science was largely a product of the nineteenth century. It stresses the relation between experiment and systematised invention; it concentrates on the creation and control of novelty. My technique is to build on historical writings that are regarded as sound and exciting, yet which remain peripheral to our received accounts of science. Here again, I extend some of Kuhn's (essential tension) arguments, especially by including the biomedical sciences as running in parallel with physical sciences. Again, I stress that I am less concerned with origins than with innovations, i.e. the entry of inventions into social practice, their becoming part of the workings of the world. So we can afford Greek ancestors and early modern progenitors, without locating there the social innovations which belong to times and places much closer to our own.

At the end of the chapter I consider the extent to which my account of experimentalism can also be applied to invention, as a way of making which developed contemporaneously, as part of the division of labour in nineteenth-century industry. I shall argue that the parallels between experiment and invention were so close that the two might be regarded as the sides of one coin.

This finding will then provide the basis of the following chapter – on technoscience.

Meanings of experiment

Though a few philosophers of science have focused on experimentation (Hacking, 1983), and several historians and sociologists of science have written at length on particular experiments (Collins, 1992; Galison, 1987; 1997; Gooding, Pinch and Schaffer, 1989; Shapin and Schaffer, 1985), we lack a commonly agreed typology of experiments. (This is another reflection of the monism of most philosophy of science and its sociological stepchildren – the assumption that science is much the same, anytime, anyplace – or that the differences are so many and various that nothing systematic can be said about them.) But we can draw on Kuhn (1977b) and the distinctions made by early methodologists who were concerned with promoting experiment.

Many early modern 'experiments' in the classical sciences were essentially *demonstrations*. If classical sciences, especially mechanics, can be characterised as analytical in reducing complex motions to a few simple principles, then this role of experimentation is easily understandable. The demonstrator creates a simple situation in which the elementary motions can be observed free of their usual (natural) complications. It is also evident why there may be a tendency to conflate real experiments with 'thought experiments', conducted 'in the head' or 'on paper'. Such experiments, like classroom experiments, are not meant to produce anything new, nor even to allow decisions between hypotheses – they are exercises in working out analytical principles to reconstruct phenomena, in principle, from their elements. Though experimentation, even in this sense, was not a significant part of ancient planetary astronomy, it played a role in optics, hydrostatics and statics. It was important in medieval studies of local terrestrial motion as an area of quantitative analysis. Historians dispute the extent to which the experiments of Galileo are to be read as demonstrative rather than as tests of hypotheses.

Experimental histories

Kuhn was keen to distinguish this demonstrative role of experiment from the tendency he saw developing in the sixteenth and seventeenth centuries, described and encouraged by Francis Bacon. Nature was to be 'tortured', preferably by instruments, so as to give up her secrets – to produce effects not to be seen in the normal course of events. He links this tendency to the fading of the distinction between nature and artifice (as discussed in Chapter 3). If God was a craftsman, then his craft work could be taken apart or put to the test in ways that were continuous with the exertions of human craftsmen. Bacon distinguished between experiments meant to be of direct economic benefit and 'experiments of light' that were meant to illuminate the phenomena; but we can follow Kuhn in seeing most such experimentation as largely separate from experiment/demonstration in the analytical sciences (Shapin, 1994; Shapin and Schaffer, 1985). Baconian experiments were meant to feed into crafts, or to provide 'experimental histories' which were part of natural history (in the wide sense). Indeed, we may see Baconian experimentation within that early modern reformulation of natural history which 'cast out' the hermeneutic and 'cast in' the 'history of nature-worked-on-by-man'. These enlargements of the field of natural history were also related to the expansion of range made possible by telescopes, microscopes and voyages of exploration. However, the field so recast and expanded remained natural history in as much as the phenomena could only be described and catalogued.

In these respects, my classification seems in line with the detailed analyses of early modern experimentation published by Shapin and Schaffer to which I have briefly referred in earlier chapters. They make the connection between the air pump and the new optical instruments as ways of making phenomena visible; they show how Robert Boyle avoided mathematics and linked his experiments to natural history; above all, they stress the production of 'facts' by public witnessing and reporting. As several scholars have argued, the status of 'facts' was also crucial for the new natural history (e.g. Daston and Park, 1998). There is then ample mandate for seeing the air pump as a contribution to 'experimental history'.

But, because I do not see such 'experiments' and natural history as the only activities characteristic of science in that period, there is also room for other readings. The dynamics of mathematics, or more generally of the analytical sciences such as planetary astronomy and mechanics, seem obviously different from those of natural history, as I have briefly explored in Chapter 4. To the extent that Boyle's studies of air involved the testing of hypotheses about the 'mechanics' of air – its 'spring' and pressure-to-volume ratios, they might be said to be grounded in an analytical reduction of 'air', and to be comparable to analytical experimentation in mechanics. Such experiments could be demonstrative, and/or measurements to establish relationships – like Galileo rolling balls down inclined planes as a relatively easy way of 'timing' falling bodies.

So perhaps it would be useful to see the air pump experiments in *several* perspectives – as natural-historical *and* analytical, and maybe also with some characteristics of the 'synthetic experimentation' on which this chapter concentrates. My suggestion is that most early modern experimentalism was subordinate to natural history, or to the limited analytical sciences. My guess is that experimentalism could only become a major way of knowing when 'synthesis of elements' became a way of creating and controlling novelties. In that sense, nineteenth-century experimentalism was built on the successes of analysis from the end of the eighteenth.

Experimentation and the age of analysis

In discussing analysis in the decades around 1800, we have seen the importance of quantification, especially in the formation of new chemical and physical disciplines. Weight was central to Lavoisier's chemistry, including the establishment of the combining-weights for each of the elements. Such measurements required considerable technical skill and equipment – both the torture instruments of the chemist's art (furnaces, condensers, etc.) and the instruments of measurement, such as balances, which had been more associated with mathematical demonstrations or perhaps with the commercial side of chemistry. Similarly, mathematicians, philosophers and engineers who analysed such physical phenomena as the flow of heat, used sophisticated thermometers and elaborate ways of protecting the 'model' from variations in external temperature (and the body heat of the experimenter, etc.). Reconstructions of J. P. Joule's paddle-wheel experiments have shown the difficulty of the task, and the way Joule drew on his 'domestic' advantages – his familiarity with thermometers used in brewing (as methods were standardised), and a cellar where the temperature was more or less constant throughout the day (Sibum, 1995). The themes of accurate measurement and protected conditions have also been well explored for academic physics laboratories in the nineteenth century. Graeme Gooday has shown how teachers of physics in the then new universities of industrial England (and in Cambridge and Glasgow) stressed accurate measurement as a discipline on which students could draw in their later (industrial) lives (Gooday, 1990; 1991a; 1995).

Of the biomedical sciences c.1900, physiology was the discipline most associated with experimentation, and historians of that discipline have often drawn a fairly direct line back from the twentieth century to Claude Bernard (c.1860), to François Magendie (c.1820), to Xavier Bichat (c.1800) and to Albrecht von Haller (c.1760). But here too, much of the experimentation or practical work might better be seen as forms of analysis and measurement. Haller attempted dynamic characterisation of body parts in terms of their 'sensibility' and 'contractility'. With Bichat, as noted in the previous chapter, this kind of dynamic classification was extended over a range of tissues/systems explicitly presented as the elements of the body. This 'general physiology' continued to be a major part of physiology in the nineteenth century, including analyses of reactivity (irritability, etc.) as well as histological and chemical analyses of body tissues (Lesch, 1984). Likewise, much of the 'physical physiology', in the new laboratories of German universities c.1840 could be seen as continuing this tradition of 'characterisation of elements', e.g. electrical studies of muscle and nerve. We easily forget that the first physiology institutes were much concerned with analytical microscopy or physiological chemistry; indeed, the wish of medical students to acquire facility in these techniques of medical analysis was one of the selling points of the new physiology (Kremer, 1992; Tuchman, 1993). Such techniques may be seen as analogous (and continuous) with the analytic and measurement skills promoted by chemical and physical institutions.

I do not claim that all early nineteenth-century physiology was analytical, but it was chiefly after mid-century, especially through Claude Bernard's writings in the 1860s, that experimentalism as a method was clearly formulated. Curiously (and this fact seems to have escaped previous commentators), Bernard's innovations closely followed the corresponding reflections of his friend and colleague at the Collège de France, the chemist Marcelin Berthelot (Crosland, 1970). To illuminate this nexus, and the much wider questions of self-conscious experimentalism, we turn to consider the methodologies of nineteenth-century chemistry, then of biological sciences and then of 'physics'.

Synthesis in chemistry

I have already suggested that synthesis in chemistry was the analogue of experimentalism in biomedical sciences. For me, initially, this analogy was a hypothesis: I was looking for analogues of Bernardian experimentalism; I found my hypothesis confirmed in interesting ways for French science, and then I learned something of the larger history of 'synthesis' in chemistry. Though historians of chemistry had written about synthesis, it

was usually as a prelude to the fine chemical industry, or in relation to the once vexed question of how the characteristic molecules of *living* bodies were related to the simpler substances of
which they were composed. But synthesis was also important as
a scientific method, a way of knowing. As Berthelot repeatedly
argued, synthesis took chemistry beyond mere analysis; it
afforded a means of checking analysis, but also of testing predictive hypotheses; and it also allowed the creation of existing
useful chemicals, or of chemicals unknown in nature (Berthelot,
1879). Berthelot was a major proponent of chemistry applied to
industry.

However, the vogue for synthesis predated the discussions in
France. It was in industrial Britain that chemistry was first promoted for its synthetic possibilities – but the Britons were led by
a German. Wilhelm Hofmann, Liebig's favourite disciple, was
the director of the Royal College of Chemistry, founded in
London in 1845 and state funded from 1853. He promoted this
pioneering institution not just by training analysts, but by pointing to the possibilities of synthesising new materials.
Interestingly, atomic models were central to his demonstrations;
wooden balls joined by sticks allowed chemists to envisage how
atoms might be arranged in compounds. If you could model
natural compounds, you might extend your models to new compounds and then try to make them. Models encouraged ideas of
'building' new molecules. It was all very practical, concrete and
British: the models were based on the atomic theory pioneered
by John Dalton at the opening of the century. Amusingly, when
Hofmann returned to Germany as a university professor, he
dropped modelling as unsuited to the 'higher', more philosophical approaches to science (Meinel, forthcoming).

Berthelot, by contrast, had never been fond of models.
Philosophically, like Claude Bernard, he was a follower of
Auguste Comte, the founder of positivism. For positivists,
science was the discovery of laws (not crude models of structures
that could not be known directly). But positivism, too, was easily
compatible with an enthusiasm for synthesis as a way of con-
firming hypotheses; indeed, for many positivists, the ability to

manipulate nature predictably was good evidence of the correctness of their scientific principles. As we shall see in the case of the biologist, Jacques Loeb, a positivistic, 'engineering' approach to science could mean a heavy stress on experiment as the highest mode of investigating life (Pauly, 1987).

It seems significant that both Berthelot and Bernard worked at the Collège de France, a privileged Parisian facility for research and special lecture courses, and *not* for the training of practitioners; both argued for the utility of their disciplines by demonstrations of experimental synthesis and control, rather than by analysing economic or medical processes. And the other famous professor at the college was none other than Louis Pasteur, who in the 1860s became widely known for his research on the causes of fermentation; his control experiments using swan-necked flasks, with which I introduced this chapter, were to become classics of 'scientific method'. But Berthelot had yet another reason for stressing 'synthesis'. He wrote a history of chemistry which featured Lavoisier as the proponent of analysis, thus making dialectical space for Berthelot himself to appear as the chief proponent of chemistry's next stage (Crosland, 1970).

Experimentation in biomedical sciences

What Berthelot claimed for synthesis in chemistry, Claude Bernard claimed for controlled experiments in physiology: they were the means of moving biological sciences beyond the merely analytical to a higher level of scientific method, the means of testing predictive hypotheses. Most commentators have stressed the obvious: that Bernard was pleading the cause of his own discipline, physiology, which he presented as 'experimental medicine'. They have noted his denigration of existing medical sciences as being *'merely descriptive'*, including statistical studies in medicine – the 'numerical method' mentioned in my previous chapter. Bernard's criticism often surprises present-day readers, because statistics are now accepted as an essential aspect of experimentalism, but he was convinced that one well-controlled experiment could decide issues in a way that transcended statistical observations (Canguilhem, 1975; Grmek, 1970; Holmes, 1974).

Commentators have also noted Bernard's attachment to the positivism of Auguste Comte, including the characteristic advocacy of biology as a general science of life, and some have stressed the multiple meanings of 'control' in Bernard's work (Figlio, 1977). Controlled experiment in his new, technical sense, involved 'controls' – systems that were just like the real experiment except for the key factor under investigation. But this was also a way of 'controlling' the experiment in a more general sense, a means of gaining control over living processes. The promise of Bernard's *Experimental Medicine* (see Bernard, 1957) was clear and persuasive: what could be controlled in laboratory animals today might be controlled in human patients tomorrow. Animal experiments would lead to a dynamic understanding of living processes, and on to therapies. And so it proved – after about fifty years of work.

There is room for more investigation of the Parisian context of these arguments. We know that some zoologists objected to Bernard's 'put down' of their science; one of them argued that he could test hypotheses by searching out 'natural experiments' where the operation of the key factor could be observed (Churchill, 1973). In an earlier debate, M. E. Chevreul, the Natural History Museum chemist, famous for his work on colours and on painting, had pointed out the role of '*a posteriori*' experiments in testing hypotheses. For example, if you wanted to know why dyes had different effects depending on the composition of the water, you could analyse the waters and guess the key factors; but then you should test the hypothesis by adding the critical factor to distilled water to see if you got the expected result. The effects of spa waters on human ailments could, in principle, be explored by similar methods (Churchill, 1973; and see Chevreul, 1824). That such matters were discussed in the Paris Academy of Sciences may suggest that moving beyond chemical analysis was not so obvious as we may now presume.

Whatever the background, we know that Bernard's work

proved influential. No one claims that he invented experimentalism in physiology, but he gave a self-consciousness to experimentalists which proved powerful, not least in its extension to other biological sciences. Some botanists, especially in Germany, adopted similar methods for plant physiology. Famously (eggs again), some embryologists declared that the future of their science lay in experiments to control development, not just to elucidate the potentialities of the different 'elements' (germ-layers) of the embryo (Churchill, 1973; Gilbert, 1994). Jacques Loeb, who moved from Germany to the USA and there helped develop experimental studies on protozoa, was one of the theorists of experimentalism. As a positivist (and follower of the physicist Ernst Mach) he took an operationalist view of science – the ability to control was the measure of knowledge (Pauly, 1987). Indeed, from the late nineteenth century, so powerful was 'experiment' as a banner that much work was so labelled which might more accurately have been described as microscopical analysis.

That elision was evident in Britain and the USA from about 1870 as T. H. Huxley and his protégés campaigned to introduce 'biology' into universities. Huxley, 'Darwin's bulldog', was a classical zoologist and palaeontologist, known for his morphological studies, but he was also an ardent advocate of a wide-ranging science of form and function that was sometimes called 'experimental biology'. He arranged that schoolteachers should be trained to teach it, that medical students should learn and practise 'experimental physiology', and that the 'new botany' should be introduced into departments previously devoted to plant classification. Though the 'unification' of biology was rarely achieved in British universities, the movement for 'experiment' succeeded in as much as it drew on strong contemporary movements for the reform of medical education and for science in elementary education (Desmond, 1994; 1997; Geison, 1978; Gooday, 1991b; Thomason, 1987). From the end of the century it grew rapidly when government began to see 'science' as the key to the development of the Empire. Tropical medicine became a British specialism – parasitology, economic entomology, mycology, etc., took shape in this context and continued to be strongly supported, for 'home' as well as imperial reasons, through to the Second World War (Worboys, 1976; 1998).

These new biological sciences were experimental in a rather catholic sense. In the UK and the USA, proponents of the 'new botany' or 'new biology' or 'experimental zoology' often saw themselves in opposition to older naturalists – amateurs or museum curators whom they tended to dismiss as merely observational or taxonomic, mere 'stamp-collectors' (Maienschein, 1991). In fact, much of the new biology was microscopical observation or analysis, e.g. working out the life cycles of fungi – the stuff of classical zoology and botany in the German universities. Experiments in the Bernardian sense were a small part of the repertoire, but they were played up. If you stressed the manipulations involved in dissections or microscopical studies, these too could be encompassed as 'experimental'.

The development of laboratory biology was roughly contemporaneous with the founding of university laboratories for experimental physics and then engineering (Fox and Guagnini, 1998; 1999). In all cases, I suggest, these sites chiefly provided protected environments for the development of *analytical* skills (and methods); they were a way of bringing some of the world into the college, but that was easier if you referred to the spaces as 'laboratories' rather than workshops, and stressed 'experiment' rather than just analysis (Fox and Guagnini, 1999: ch 3). The laboratories were chiefly for teaching, and experimentation in a strict sense was a relatively small part of the work, but it was ideologically and prospectively important because it established university laboratories as creating novelties rather than simply analysing the world outside. We see here, again, the politics of ways of knowing and the power relations between them.

Experimentation in physical sciences

Historians who have studied physics have usually focused on physical theory and especially on generalisations such as the doctrine of energy, briefly discussed in Chapter 4. Latterly some have given detailed studies of experimentation, especially in the seventeenth and twentieth centuries (e.g. Collins, 1992; Galison, 1987; 1997; Gooding, 1985; Shapin, 1994; Shapin and Schaffer, 1985). Yet we lack a general history of 'experimental' physics, a term in common use in the later nineteenth century to contrast with mathematical physics. Historical studies of measurement cover part of this field, but in a sense they serve to highlight the key question for this chapter: what was distinctive about physical experimentation, *beyond measurement*? How did the methods and methodological statements of 'physicists' relate to those of contemporary chemists and biologists? Is it fruitful for us as historians to look for analogues of synthesis in chemistry or Bernardian experimentation in biomedicine?

That such questions seem scarcely to have been asked reflects two features of most modern history of science, against both of which this book is directed. First, the disciplinary divides among the historians who deal seriously with the intellectual content of the sciences and, secondly, the presumed primacy of physical sciences – usually justified in terms of mathematical sophistication, theoretical rigour and experimental accuracy. Philosophers of 'science' have usually read 'science' as meaning physics, and historians of science have usually seen chemistry and biology etc. as trailing along behind 'physics'. The classical post-Second World War history of science by Herbert Butterfield is exemplary here. It focuses on the scientific revolution of the seventeenth century which was mostly 'physics', the chemical revolution was delayed until the end of the eighteenth century, and biology brought up the rear with the Darwinian revolution of 1859 (Butterfield, 1957). Note, too, how the landmarks are changes of theory rather than of method.

This story of hierarchy and of philosophical and historical precedence has some merit; certainly its prevalence needs to be explained, and we need historical explorations of the extent to which this hierarchy was felt by different kinds of researchers at different periods. But we also need research from the opposite supposition – that the various sciences at any given period shared questions of method and methodology; that in some ways they proceeded in parallel. This 'hypothesis of synchronicity' seems especially promising for the nineteenth century, after many new analytical disciplines had taken shape together. Whether we look to Paris as the world centre of the analytical sciences, or to London and the work of Humphry Davy and Michael Faraday at the Royal Institution, we have a sense of analytical researches developing in parallel across a broad front, with no clear differentiation between physics and chemistry. Many different elements, including light, electricity and heat were under investigation (Knight, 1986; 1992).

This specific feature of the time is clear in the contemporary discussions of the 'classification of the sciences' (Fisher, 1990) and in that formative text on 'scientific method', John Herschel's *Preliminary Discourse on the Study of Natural Philosophy*, first published in 1830. The son of William Herschel, the astronomer who classified stars, John was a major protagonist for 'science' and for 'scientists', as some of his friends had begun to call themselves. As an admirer of Francis Bacon he did much to create a self-conscious experimentalist tradition in England, but he was also very clear about the role of analysis in the constitution of the sciences around him: 'In pursuing the analysis of any phenomenon, the moment we find ourselves stopped by one of which we perceive no further analysis, and which therefore, we are forced to refer (at least provisionally) to the class of ultimate facts, and to regard as elementary, the study of that phenomenon and of its laws becomes a separate branch of science' (Herschel, 1835: 93).

From the middle of the century it is easier to talk of 'physics', when elements such as heat and motion were covered by more extensive notions such as energy, and as more interactions between elements were uncovered. In German universities, *Institutes* for physics accompanied or followed those for physiology; chemists usually led the way. In late nineteenth-century industrial Britain, the development of university departments usually followed this same order but, in Cambridge, chemistry followed the other two. In all these sites, 'experimental physics' was among the new disciplines created by the nineteenth century, but it was usually experimental in the wide sense discussed above for biology – much of the work was analytical.

'Natural philosophers' were indeed much concerned with measurement of these elements such as light and heat, but some 'experimental philosophers' also focused on *interactions or reactions* between elements – the chemical elements or heat, light and, especially, the various kinds of electricity. Here, perhaps, one can see the beginnings of experimentalist traditions that were to produce 'the new physics' of radioactivity etc. at the end of the century. They were marked off from the Baconian experimentation of the eighteenth century by the focus on new 'elements'; they were both analogous to and continuous with experimentation in chemistry (and engineering). For Britain, the founding father of this tradition was Michael Faraday, the patron saint of later experimentalists (Cantor, 1991; Gooding and James, 1985; Williams, 1971, 1987).

Faraday, the son of a blacksmith, became assistant to Humphry Davy at the Royal Institution (in some ways the London equivalent of the Collège de France). There he acquired the skills of a chemist, and followed up Davy's work on electrochemistry. Faraday was no mathematician and he was not primarily interested in measurement. He did lots of analytical work, not least for the government, but his public work centred on searching out possible interactions between 'elements' (partly because of a metaphysical belief that they were but different expressions of a common force).

Historians of physics tend to stress this metaphysics and the way which he 'intuitively anticipated' field theory. Chemists have found Faraday's style more natural, because of his cataloguing of experiments. Here it may be instructive to recall how Faraday systematically checked out the electrical properties of many materials (paramagnetic and diamagnetic, insulators and conductors). At one stage he worked through the various 'kinds' of electricity (electrostatic, voltaic, magnetic, thermal, fishy, etc.) to establish their identity. His most famous works concerned the interactions, assumed to be reversible, between the various elementary forces – magnetism and current electricity, electrical forces and motion, electricity and polarised light, electricity and crystalline structure, light and electrochemistry, etc. He discovered many such interactions or 'effects', and he 'prepared' the discovery of others, accomplished later by researchers who employed more powerful equipment or happened to use the particular materials which 'worked'. I suspect that much of the experimental physics of the nineteenth century could be expressed in a matrix of real and possible interactions between these 'elements' or 'forces'. Among Faraday's contemporaries and friends one sees a similar pattern in the work of Gustav Magnus, who in the laboratory at his home created the Berlin tradition of experimental physics (Fox and Guagnini, 1998; Kauffman, 1974).

Faraday's successor at the Royal Institution was John Tyndall, a pugnacious ally of T. H. Huxley in their many struggles to secure more resources for scientists and less authority for bishops (Desmond, 1994; 1997; Turner, 1980). Tyndall, again, was an experimentalist; he had worked with Magnus in Berlin, and he ranged widely across the (analytical) disciplines. He followed up Faraday's work on magnetic properties of various materials, involving the compression of hundreds of crystalline substances. These studies led to research on compression of ice (and glacier movement) and on the effects of solar radiation and heat radiation on atmospheric gases; hence to the scattering of the sun's rays on dust, to the amount of dust in atmospheres, to floating organic matter and to bacteriology (Macleod, 1976). We note here the preoccupation with 'outdoor physics' or the physics of 'nature'. We shall return to dust, as it were, and to nature, but first a glimpse of Paris and the French tradition.

In 1838 a Chair of Physics was created at the Museum of Natural History, which we have already visited for zoology, botany, geology, crystallography and chemistry. For the rest of the century, the chair of physics was held by three successive members of the Becquerel family. Antoine (1788–1878), for whom the chair was created, worked on diamagnetism before Faraday; he investigated the generation of electricity by pressure on crystals, and by heating minerals; he also experimented on the synthesis of new crystalline forms of electric fields. From the 1870s the laboratory received state support for the development of electrical studies (Fox and Guagnini, 1998: 111–12). His son, Alexandre, became a world expert on the relationships between light and chemical reactions. His grandson, Henri, did further work on the effects of magnetism on polarised light, and on crystals which absorb or emit light; in 1896 he used X-rays on luminescent crystals – and thereby discovered radioactivity, the basis for an unexpected expansion of experimental physics in the twentieth century.[2]

We note here the continuity between this experimental physics and other studies at the museum, especially crystallography. We see how such studies of 'effects' could illuminate general theories (e.g. of the relationships between light and electricity), but that they were also investigations of natural curiosities (such as photo-luminescence), and the basis of useful instruments (such as the actinometer which used chemical reactions to measure quantities of light). We also see how 'the new physics' of radioactivity sprang from this long tradition of investigating 'reactions between elements' – experiments which 'bridged' chemistry and physics. We can see another aspect of that bridging if we examine the physics of clouds. Here I take my lead from an article by two American historians of modern physics, Peter Galison and Alexi Assmus (1989), but I develop the argument in terms of 'ways of knowing'.

On clouds, dust and control

At the start of the nineteenth century there was much interest in the forms of clouds, not least among landscape artists such as John Constable. (And at the end of the century, John Ruskin continued the tradition.) We could call this 'morphology', in a loose sense, but it was not much more than a taxonomy of appearances, a matter of natural history. To *explain* the conditions under which clouds formed you needed to know about the solubility of water in air at various pressures and temperatures, but this *analytical* understanding became so commonplace among chemists and other investigators that *we* hardly notice it. Increasing experience showed, however, that the simple analytical framework was insufficient – as gases were cooled or pressures lowered, water did not condense out unless particles were present around which droplets could form and produce 'clouds'.[3]

Cloud formation was studied by John Aitken, a Scot trained in engineering, who devoted himself to experiments in natural philosophy – one of the many unpaid 'devotees', like Joule and Darwin, who produced much of Britain's science in the nineteenth century. The study was continued by C. T. R. Wilson at the Cavendish Laboratory, founded in Cambridge in the 1870s as a home for measurement and for experiment (Schaffer, 1992; Sviedrys, 1970; 1976). Wilson came from the Lake District and was fascinated by meteorological phenomena, which he modelled in the laboratory. When he later discovered that ions could also act as centres of droplet formation, his 'cloud chamber' became an instrument for investigating charged particles – the key tool in the Cambridge investigations of particle physics. Wilson continued to focus on the *analytical* uses of the cloud chamber, while his more famous colleagues built experiments around it, and so helped build atomic physics. Galison (1997) has explored at length the continuing interactions and tensions between analysts, experimenters (and theorists) in twentieth-century physics.

But to grasp the shape of nineteenth-century physics we must not rush to the twentieth, rather we should ask about the ways of experiment and their relations to other ways of knowing and working. Here we note the recurrence of 'dust', already mentioned in connection with Louis Pasteur and John Tyndall. Aitken and Wilson also performed classic 'dust' experiments of the kind that Chevreul called *a posteriori* and Bernard called control experiments. All four, independently investigated 'air' from which dust had been removed, and then the effects of 'adding dust' – whether on cloud formation or on the fermentation of organic matter. These 'synthetic' experiments confirmed the role of 'dust' in these various phenomena – they confirmed the significance of a particular 'element' in the analytic understanding of these phenomena. The authors would have recognised their method in John Herschel's Baconian text, to which we referred above: 'We make an experiment of the crucial kind when we form combinations, and put into action causes from which some particular one shall be deliberately excluded, and some other purposedly admitted; and by agreement or disagreement of the resulting phenomena with those of the class under examination, we decide our judgement' (Herschel, 1835: 151).

Adequate *analysis* of 'causes' allowed the widespread development of experimentalism. As Faraday noted: 'As an experimentalist I feel bound to let experiment guide me into any train of thought which it may justify; being satisfied that experiment, like analysis, must lead to strict truth if rightly interpreted; and believing also that it is in its nature far more suggestive of new trains of thought and new conditions of natural power' (Thompson, 1898: 242).

But we note two further characteristics of these experiments on clouds that bring them alongside the chemical syntheses and the biological experiments in a way that seems stranger for physics because of the usual emphasis on abstraction, theory and mathematics. The cloud experiments could be models of real clouds; they were 'mimetic', to use Galison's term. (Would one need a term for chemical or biological experiments where mimesis seems more natural?) One way of using a cloud chamber was to play with it to reproduce natural phenomena – that aspect fascinated Aitken, and Wilson never lost his interest in meteorology. Like chemists 're-creating' an organic compound, or physiologists inducing diabetes in an experimental animal, physicists were modelling 'nature' – with the potential for control.

This kind of experimentalism was probably more widespread than most histories of physics suggest. If we look for case histories in Manchester *c*.1880, in the first of the English provincial universities, we find the academic engineer, Osborne Reynolds, very much interested in 'outdoor physics'. He is known for his analyses of flow and lubrication (Reynolds's numbers), but he also modelled river estuaries, and (for fun?) he showed how it was that lightning could explode trees (by boiling the sap). His chemist colleague, Henry Roscoe, and the physicist, Arthur Schuster, were experts on spectroscopy, reproducing in laboratories spectroscopic patterns resembling those observed from heavenly bodies (Schaffer, 1995). The techniques used in the analysis of natural or technological phenomena, could also be used in experimental re-creations, but the experiments could also become analytical techniques – in some contexts, for some purposes, experiments could be subordinated to analysis (Davies, 1983; Kargon, 1977).

Here we recall the remarks in my first chapter about the nesting of 'ways of knowing', and the different possible hierarchies. We are now in a position to expand that discussion and to reflect on the (contested) divisions of the sciences, on differences in status between such branches, and on the ways that such status differentials have varied over time and between sites. These reflections will also serve to introduce the final section of the chapter – on the relations between experiment and invention.

Experimentalism and hierarchies of knowledge

We noted that Alexandre Becquerel, by studying the effects of light on chemical reactions, created an actinometer that could serve to measure light. Tyndall's work on dust in air, and the work on cloud formation, gave rise to instruments for measuring the quantity of dust in air (and thus to the investigation of fogs, smogs and airborne infections – major problems for Victorian cities). Experiment-as-analysis is a key part of the analytic-synthetic dialectic, and of the relation between laboratory studies and investigations of 'the real world'. These problematic relations were formulated in different ways in different disciplines, reflecting the different power relations.

Sometimes studies of 'outdoor physics' or of 'technological

systems' (and the associated instrumentation) have been known as 'applied physics' (and downgraded thereby), but perhaps we would do better to use another terminological frame, common in medical sciences and chemistry – the contrast of 'general' and 'special', e.g. general pathology and special pathology, where 'general' covers such processes as inflammation, and 'special' refers to particular diseases or the lesions of particular organs. Chemists also referred to general chemistry, e.g. the study of reaction mechanisms, and to special chemistry, e.g. of chlorine compounds. In these formulations the contrast is not between pure and applied, but between a push to general explanation and a concern with explaining the complexities of the world; between, in my terms, the study of the properties of 'elements' and the properties of particular 'compounds or systems'. We do well to remember how much 'physical science' could have been called 'special physics' because it was about 'compounds'. And so it remains.

Of course, the distinction between elements and compounds is relative. Part of the progressiveness of science is the potential to envisage 'elements' as 'compounds' (or as derived from a fundamental element), for example to see atoms as 'planetary systems' of 'subatomic particles'. The 'higher destiny' of the cloud chamber was to be a detector for charged particles, a means of analysing the disintegration of atoms (and a means of establishing that disintegrating atoms also existed 'in nature'). So analysis and synthesis could be reconstructed in dialectical relation at a deeper level, and the 'general' to 'special' fulcrum shifted accordingly.

So consider the remark of Rutherford, Schuster's successor as Professor of Physics in Manchester – that science was either physics or stamp-collecting. He was claiming hegemony not only for experimentalism over (mere) analysis and natural history, but also for the pursuit of the general over that of the 'special', and for his new physics over the electrotechnics which occupied much of 'his' building in Manchester. As we shall see in the next chapter, Schuster had regarded the electrotechnics as vital and had argued for continuities between workshops and university laboratories, but Rutherford was not interested; the electrical work could be done in departments of *electrical engineering* in universities or in technical colleges (as was indeed happening in Manchester) (Wilson, 1983). For him, the constituents of atoms were the centre of 'science', an attitude which remained common for most of the twentieth century. It was reinforced by the link between atomic physics and astronomy, and thus with explanations of the infinitely vast, powerful and primordial. After the Second World War, the atomic physics which had once served as paradigmatically 'useless', promised cheap, clean energy and military dominance – as well as being fundamental. This view of fundamentals, when sold to governments by a generation of physicists bound together by their war work, created huge international facilities for experimental physics, e.g. CERN at Geneva. Now, however, as this vision has come to be rivalled by visions more biological and commercial, we can see more clearly how it was constructed, and how and where experimentalism was contested.

For example, as we have seen, one of the key sites for experiment was physiological laboratories of the later nineteenth century, where chemical and histological analysis was mostly subordinate to the experiments on live animals conducted by the head of the physiology institute. But suppose the head of the institute was a chemist or a histologist, rather than an exponent of vivisection. Under such circumstances the focus might be analytical and possibly experiments might be undertaken to produce new substances for analysis, e.g. to complete an analytical series. Or, when biological substances must be measured (and standardised) in the absence of chemical tests, researchers resort to routine 'experiments' called bioassays – say to measure the effects of 'nerve-growth factor' on the spinal ganglia taken from early embryo chickens. In such bioassays, the aim is to measure the substance, not to explore the effect; 'experiment' here serves analysis.

More generally, in clinical contexts such as hospital laboratories, the analytical values of diagnosis and therapy may be predominant and experimentalism may be subordinated. In early

nineteenth-century France, the Paris Academy of Medicine promoted physiological researches on the mechanisms of vomiting
in animals. In a university laboratory these might have been a
means of exploring reflexes, for clinicians they were orientated
towards elucidating the conditions that provoked vomiting in
patients (Lesch, 1988). Similarly, electrical stimulation of nerve
and muscle could be a means of exploring normal and abnor-
mal reaction mechanisms, or a means of clarifying neurological
and muscular disorders. Clearly these orientations are matters
of degree, but subtle differences of this kind are often important
in the politics of knowledge and in determining the development
of research in particular sites. The history of 'clinical research'
or 'research in engineering' suggests that in institutions oriented
to practice, analysis has often been the preferred mode of
science, and in such situations 'experimentalism' may be subordinated.

The details do not matter much, but the general point is important. In discussing scientific projects we can analyse the component ways of knowing and the logical relations between them,
but we also need to assess the 'direction' of the project – its leadership and its orientation in terms of the predominant way of
knowing. If we seek an explanation for the overall form, then we
must look to the contingencies of social life and to the politics of
knowledge as favouring certain modes over others. Much the
same applies to the relations of experiment and invention.

Experiment and invention

In preparing to define experimentalism in my specific sense of
'synthesis', I also related experiment to natural history (experimental histories) and to analysis (analytical measurements, etc.).
Similarly, when we try to pin down the specificities of invention
as a major function of nineteenth-century technology, we need to
consider a comparably wide range of meanings, where invention
appears alongside the other ways of making (and knowing) –
alongside craft (and natural history) and rationalised production

(and analysis). I want to suggest that crafts are generally inherited rather than novel, and that rationalised production in some sense reconstitutes existing processes, but that (synthetic) invention like (synthetic) experimentalism creates novelties – things and processes that did not exist before. That is the core of my usage, though clearly, definitions are slippery.

Some 'inventions' may be regarded as 'findings' – from '*invenire*', 'to light upon', a usage in early modern English.[4] Roast pork may have been invented (or discovered) in a burnt pig-sty, some medicines were discovered by chance. By 'torturing' materials, in Francis Bacon's sense, we may find new and useful properties, e.g. new kinds of hardened pots or better curing of leather. Here we see the links with natural history and craft, and in as much as more sophisticated inventions comprised many minor adjustments attained by trial and error, so this simplest kind of invention may indeed be fundamental (Basalla, 1988). But usually invention has involved some sort of *analysis*, some attempt to assess a 'need' or an economic possibility. Why, then, do I *contrast* invention with analysis and rationalised production? Was not James Hargreaves's spinning jenny an 'invention' (see O'Brien, Griffiths and Hunt, 1996)?

The spinning jenny modelled the actions of spinning on a wheel, it mechanised the twisting and the backward-and-forward motions, and allowed one person to work several threads simultaneously. One suspects that many people had thought of 'multiplying' spinning in this way, but doing so successfully involved craft skills and persistence. We could call it an *analytical invention*; the industrial world is full of them, because so many of its commodities are mechanically produced substitutes for prior craft products. Of course, you can argue as to when analytical inventions move beyond 'reconstruction' or multiple replication and so produce 'novelty'. The mills of Manchester in 1800, seven-storey palaces packed with machines, were novel in many senses; they involved much ingenuity, and called for management (another novelty); but still they mostly reconstituted craft processes.

Yet clearly, in some contexts, inventions can be *synthetic* novelties, intended to meet aims imagined but never realised. Many such goals, e.g. human flight, persisted for centuries before they were attained. Like synthetic experiments, invention may be led by hypotheses about new worlds, or worlds where old ends required new means.

'Invention is a new combination from "the prior art"' says a curious book on the sociology of invention, full of aphorisms, most of which seem rather obvious (Gilfillan, 1962). But they will serve here to stress the idea of combinations or synthesis, at the level of materials or processes, resources and purposes, as in the curious case of barbed wire. It was invented in the great plains of North America as they were developed for cattle ranches; fences were needed to restrain the movement of cattle, but wood and stone were rare and the lengths of fence required were great. Two methods were tried: one was traditional in other places – thorn hedges; the other was an adaptation of a new industrial product – (plain) wire made into fences. Neither worked well, but then someone thought of the hybrid. Wire fences with wire 'thorns' worked (and sold) very well indeed (Basalla, 1988: ch. 2).

Such examples show a strong analogy between synthetic invention and synthetic experiment, but obviously there are many differences other than of overall purpose. The components of barbed wire were everyday objects, not the objects of detailed analysis or any other form of close study; it seems natural to describe it as an invention, and its initial use as a 'trial' rather than an experiment. But of course, the relations may change over time. Inventions, like any other technological artefacts, may prompt analysis for the rationalisation of production and use, or indeed raise problems that attract investigators who are building systematic knowledges. No doubt the compositions and dimensions of barbed wires have now been closely analysed to effect the greatest economy in meeting any particular use, and perhaps the form of the barbs has been refined from studies of the distribution of nerve endings in bovine skin. (Who knows – perhaps there are technoscientific networks for 'bovine-kinetics'?)

But if synthetic invention may prompt analytical (or experimental) researches, what shall we say of the converse relation? What can we say about analytical studies as (re)sources for synthetic invention?

Perhaps we have already given the core of an answer in our discussion of analysis and experimentalism. The elements discovered by analytical sciences were *generalisable* – that is what made them 'elements' for science, not just 'elements' of a particular craft process such as weaving. So these elements could be put together in new ways – not just as synthetic experiments (for light), but also as 'synthetic' inventions (for fruit). The massive extension of analytical STM from about 1800 made available many new procedures for handling 'elements', which were of use to inventors; and the experimentalism we have discussed, from Faraday through to late nineteenth-century universities, produced 'effects' and equipment, such as model electric motors, which inventors could play with and develop for economic uses. Many of the classic electrical inventions of the later nineteenth century were derived from laboratory devices – put together in various ways to produce 'economic' effects. In such cases, we begin to see, not only the similarities between synthetic experiments and inventions, but their reciprocal, creative and generative relations. Model systems that could produce both light and fruit formed the core of new (synthetic) technologies.

These later nineteenth-century symbioses between science and technology did not mean that all inventors were graduate scientists, but they suggest why up-to-date intelligence from university laboratories eventually became a desirable commodity, and why some academics also benefited from the interaction. Collaboration between universities and industry was formative for a few areas of STM at the end of the nineteenth century, notably in electrotechnics and synthetic dyestuffs-cum-pharmaceuticals; in the twentieth century it became crucial to universities, companies and governments. The growth and varieties of these interactions, but also their limits, are the subject of the next chapter.

Notes

1 On Pasteur, see Geison (1970–80; 1995), Latour (1987a) and Saloman-Bayet (1986).
2 For the various Becquerels see the *Dictionary of Scientific Biography*.
3 The conditions of formation of dew, as investigated by Dr Wells, was one of John Herschel's key examples of scientific investigation. See Herschel (1835).
4 For stimulation on the relations of experiment and invention I am indebted to a (then?) unpublished paper by Johann Martin Stoyva.

7

Industries, universities and the technoscientific complexes

Let's first survey the present relations between universities and commercial companies in high-tech industries. From the universities, the companies receive graduate scientists for employment at various levels, ranging from first-degree holders through trained researchers to leading experts capable of establishing major new industrial research programmes in the company's laboratories. They may well also receive formal or informal advice from academic consultants who draw on personal knowledge and experience, or perhaps on the work of their university research groups. The company library contains scientific periodicals, the contents of which are mostly written by academics. Perhaps the company researchers may use university facilities from time to time, though for large companies the relation may now be reversed. From the companies the academics receive research support – perhaps grants for postgraduates to work on topics of interest to the company, perhaps a contribution to the running costs of the laboratory. Or the support may be in kind – the supply of instruments or of research materials such as rare chemicals or specially bred laboratory animals. In addition to direct support from companies, academics may benefit from government or university funds because the research seems likely to be important for industry and their students work on problems at the cutting edge of technology.

The interchange of personnel, information and resources is now very considerable; few high-tech firms and few leading departments could manage without it. But as I mentioned earlier, the networks also include other organisations: for medical sciences and technologies, hospitals are crucial, but general practitioners may also be involved – in clinical trials and more generally as purchasers of medical products. In fields with military importance, scientists and engineers employed by the armed forces are central. The extensions to agriculture are obvious. In all these fields, government laboratories may be involved as innovators; more generally they provide standards, test materials, check specifications, assess needs, etc. In the most expensive fields of science, where equipment costs more than universities or companies can afford, then governments may be persuaded to oblige – individually or collectively. Academics may thus gain access to rare facilities, and industrialists gain contracts to construct the equipment and benefit from 'spin-off' technologies. Governments gain prestige, and hope for military or industrial advantage.[1]

These are the *technoscientific* complexes which dominate the knowledge and commodity production of our time. For our analysis of them, as I suggested earlier, we gain more by recognising the dense interweavings of universities and industry, and of science and technology, than we do by trying to separate them. This chapter is about the development of the interactions, about the objects, knowledges and skills that moved between industry and academies, and about the social relations that were both the prior conditions and the products of those movements (Latour, e.g. 1987a; 1987b).

In the last chapter I introduced 'synthetic inventions' as the analogues of 'synthetic experiments' and suggested that these two kinds of synthesis were sometimes intimately related. Typically, the 'model systems' of laboratories could provide either light or fruit, to use Bacon's terms; where both philosophical and practical aspects were pursued, such models become the core of *technosciences* – of institutions and projects that produced knowledge-commodities. If we see technoscience as a way of *knowing and* a way of *making*, we underline the confluence of the scientific and the technological histories with which this chapter is concerned. We also allow for *different kinds* of technoscience. For example, major expeditions could be thought of as *natural-historical technoscience* – networks of academics, government and industry in pursuit of specimens. The networks of professional institutions and government/military services in post-Revolutionary France might usefully be called *analytical technoscience*, and early in this chapter we will explore further some of the analytical relations between academics and industry in the nineteenth century. But this chapter goes on to trace the closer forms of academic–industrial collaborations that I have called synthetic technoscience – when academics and industrialists work on model systems that are philosophically *and* commercially interesting, and when synthetic experiments/inventions are developed in networks of universities, research institutions and industrial research laboratories.

I suggested previously that such networks involved all our ways of knowing and of making – natural history and craft, analysis and rationalised production, and synthetic experiments and inventions – but we can also note that they could be variously 'directed'. Technoscience networks were not all 'industrial', some projects were 'governmental' in as much as they were led by state agencies for military, public health or national prestige reasons. Some other projects were directed by academic communities which had persuaded governments that the likely products or prestige merited public investment.

This chapter surveys the key industrial technosciences of the late nineteenth century (electrotechnics and pharmaceuticals) and then outlines some of the key state projects in the two world wars (e.g. poison gas and the atom project). The state-academic-industry linkages built in the First World War were often transitory, those of the Second World War continued to be developed post-war, in the USA and the UK. I conclude by surveying the relative decline of state-technoscience and the continued rise of industrial-academic linkages at the end of the twentieth century.

But in tracing the developing intimacy of science and technology, we must always remember that for most of the nineteenth century, and for many sectors up to the Second World War, the development of industry was not heavily dependent on universities or other forms of higher scientific education. And where it was so dependent – for chemistry in most countries, and in Germany for most technologies – the symbiosis was largely at the level of *analysis*.

Analysis and established technologies

In Chapter 4 we have already discussed some of the characteristic nineteenth-century interactions that we might call *analytical* technoscience, including the use of graduates (especially chemists) as consultants, and their emergent roles in public regulation of health and hazard. Especially in the chemical industries, many proprietors had a good knowledge of science and good connections with local lecturers and scientific societies. In mechanical and civil engineering the same networks operated – between proprietors, consultants and teachers – mostly to solve practical problems. Eaton Hodgkinson serves as a Manchester example; a 'devotee' mathematician who worked with William Fairbairn on design of girders, and on a radical new 'tubular' bridge over the Menai Straits, he later became professor of engineering at University College, London. Hodgkinson conducted experiments on strength of materials, but these were done in Fairbairn's works (Rosenberg and Vincenti, 1978). In many areas of engineering, the technical problems were solved by trial and error, or by systematic testing rather than by calculation. Joseph Whitworth, for example, developed his rifles on a testing-ground at his home in the suburbs of Manchester (Musson, 1975).

Few of the 'great (British) engineers' were graduates, but that did not mean they relied only on craft skills. Many of them had been apprenticed in large engineering works – which remained a preferred means of training until the end of the nineteenth century. Most of them understood some mathematics, read the engineering magazines, pursued patents and knew about developments elsewhere; they were often major sponsors of formal scientific education – thus enlarging both the community of experts and the supply of educated workers. The mechanics' institutes that made science available to the working classes in most industrial towns after *c.*1825 usually numbered engineers among their supporters, and the science colleges and provincial university colleges founded from the 1850s were similarly supported (Inkster, 1997; Inkster and Morrell, 1981).

Owens College, Manchester, after an uncertain start, flourished from the 1860s in part because its chemistry professor, Henry Roscoe, could count on the support of key local industrialists. A chair of engineering was created and filled by Osborne Reynolds, chiefly through the support of local engineers; by the end of the century, a university degree was a suitable prelude to a high-level apprenticeship. Joseph Whitworth gave part of his fortune to establish scholarships in engineering, and much of the rest went to the university, the technical college and the new teaching hospitals. In turn, the university professors and their graduates acted as advisers, consultants or managers for local industry (Kargon, 1977).

That was the English pattern, but Germany from the 1830s was the chief foreign model in such matters, and one that was sometimes imported. The leading chemists of mid-century Britain had mostly trained in Germany, until Hofmann (whose atomic models we discussed) extended that educational facility to London. Some German chemists worked in British industry temporarily, e.g. Heinrich Caro who returned from Manchester to lead German dyestuffs research; some stayed in Britain and built chemical businesses, often becoming advocates of technical education.

In Germany, the higher technical colleges taught engineering (Lundgren, 1990). Many of the leading engineers were educated in them and some of the graduates moved to Britain or the USA, where they pressed for the inclusion of engineering in universities. Several of the engineers in Manchester, e.g. C. F. Beyer in locomotive manufacture, and Henry Simon in flour milling, had trained in German (or Swiss) polytechnics, and were major supporters of the university and technical college (Simon, 1997). But even in 'advanced' cities like late nineteenth-century Manchester, with its strong community of science teachers and education-minded industrialists, each side knew its place.

Chemists in the new university taught analysis and developed experimental disciplines; they acted as advisers to industry, but did not themselves experiment with industrial processes. That closer interaction *might* have occurred in technical schools, in the programmes of education for specific trades which began to develop from the 1850s, but there was little state support until the 1890s, and problems of resources and of industrial secrecy and rivalry prevented much close engagement with specific technologies.

The situation was a little different in the other great capital of industrial Britain – Glasgow. As a Scottish city, Glasgow had a much stronger tradition of public education and a long-established university, including a medical school. Glasgow, as we noted, had a chair of natural philosophy to which they attracted William Thomson, an accomplished mathematician and experimentalist, much interested in engineering problems, not least the emergent electrotechnics. But Glasgow from 1838 also had a chair of engineering, the holder of which made major contributions to the analysis of steam engines (Channell, 1988). This formal support for activities which were informal in English regional capitals encouraged closer relations between academe and industry, especially around telegraphy.

By c.1900 the leading universities and colleges of Britain had laboratories where students could 'practise', and in the twentieth century most consultant engineers had formal training, at least in evening classes in technical schools (Guagnini, 1991); but we should not exaggerate the impact of higher education, or systematic research, on industry generally. Many forms of manufacture that were new(ish) c.1900, e.g. rifles, bicycles and cars, involved lots of invention but few inputs from academics. British motor manufacturers conducted little 'research' and employed few 'graduates' until after the Second World War (Buchanan, 1989). 'Developments' were usually combinations from 'prior arts'; relatively little came from 'laboratories'. Certainly that is true compared to 'electrotechnics', which became one of the key synthetic technosciences of the later nineteenth century.

Electrical analysis and synthesis

In some senses electrical technologies were born in the laboratories of natural philosophers; they were 'born into' worlds of measurement, quantification, demonstration and the search for principles.[2] The relation between electrical principle and practice was close, perhaps especially in Britain where the key discoveries of the early nineteenth century coincided with strong amateur cultures of science and technology. Faraday at the Royal Institution was the archetypal electrical experimentalist, and his audiences contained many professionals and manufacturers with practical interests (Berman, 1978). A taste for natural philosophy was well developed in middle-class and working-class circles, especially in the industrial towns; and savants, artisans and inventors concerned with steam engines might also be interested in batteries, motors and dynamos.

Faraday himself did not pursue phenomena for commercial advantage (though he spent much time on practical matters such as mine safety and lighthouses), but his close associates were ready to take up his experiments and develop their uses. When Faraday showed how moving wires within a magnetic field could generate an electric current, his friend, Charles Wheatstone, developed the effect to create the electric telegraph. (The American inventor of the telegraph, and the code, was Samuel Morse.) Wheatstone was primed to a 'need' by previous work on the transmission of sound. He worked in his family company making (innovative) musical instruments, and he recognised electricity's potential as a better way of transmitting vibrations (Dostrovsky, 1976). For the most part, telegraphy continued to be developed by inventors and engineers but, as in the case of tubular bridges, difficulties were sometimes met by resorting to systematic testing and perhaps by more sophisticated forms of analysis.

As technologies were extended, so esoteric, analytical advice might prove more necessary. Creating a transatlantic cable telegraph proved to be a major technical challenge that eventually outstripped the know-how of the engineer to whom the company had entrusted the technical direction; but William Thomson (later Lord Kelvin) had predicted from the beginning that very small currents would have to be used, requiring very sensitive detectors such as he had designed. He was proved correct and his Glasgow university laboratory proved invaluable – for instrumentation, for testing cable materials and for training telegraph employees; student laboratory exercises included real industrial tasks; part of a university department became interdigitated with engineering companies. Thomson's science rescued a very expensive, high-prestige project and it made him a fortune (Smith and Wise, 1989).

Telegraphy was a major new communications technology of great commercial and political importance, especially for imperial powers, but its impact was far surpassed by the development of electrical supply systems in the last two decades of the century, first in the USA. These were initially for electric lighting, but could also be used for electric power systems for tramways (Hughes, 1983). One of the consequences was a huge demand for electrical engineers and especially technicians. The development of technical education at the end of the century centred on this demand (which was also met by private courses). In a sense then, it was for electrical technology that formal education became routine for technicians, as it was for telegraphy that testing laboratories had become routine for engineers.

One can see some of the consequences at university level by looking again to Manchester. In 1900 a new *institute* was opened for physics and 'electrotechnics'. It contained a large laboratory filled with the latest electrical equipment – generators, motors, lighting systems, etc., on which the students could 'work'. The laboratory was closely associated with major electrical firms in the region, and it was named after John Hopkinson, an academic-engineer from a Manchester family of engineers and academics, who had been killed in a climbing accident when aged 49.

John Hopkinson had trained in science at Manchester and in mathematics at Cambridge before relinquishing his fellowship at Trinity College to work for an optical company in the West Midlands. He became interested in electricity through the influence of Thomson. After six years he moved to London as a

consulting engineer and remained in practice after appointment
to a chair in engineering at King's College, London, where he
directed a laboratory sponsored by the Siemens Company. There
he worked on alternating currents, especially the design of
dynamos, on which he also collaborated with his brother who,
again after an elaborate education at Manchester and
Cambridge, had become head of the new electrical division of
Mather and Platt, a large Manchester engineering firm best
known for textile machinery. John also served on committees
that established standards for electrical lighting (Dibner, 1972;
Fox and Guagnini, 1999: ch. 3; Guagnini, 1991). Such careers
nicely demonstrate how university training, advice and research
became important to the electrical industry, but the synergy can
also be seen from the 'other side', through developments within
industry, especially of research laboratories.

There our best guides are studies of American firms, whose
early and successful research laboratories have attracted several
good historians (Hounshell and Smith, 1988; Israel, 1992;
Reich, 1985; Wise, 1985; and see Dennis, 1987). Such laboratories are easily seen as the hearts of modern industries, as the
epitome of technoscience, but we need to be careful about dates,
and about the variety of industry. Whatever the strength and
extent of that model around the year 2000, it was rare in 1900.
I have outlined the British history first, partly because we need
to be wary of seeing science and technology as inevitably con-
verging into the industrial research laboratories pioneered by
American electrical firms and German chemical companies.
While *some* such 'convergence' might be seen as over-determined
in the long-term – as secured sooner or later in most of the relevant sites – the form and dating of developments was highly con-
tingent, and the pattern was not widely established until after the
First World War. Indeed, *several* notes of caution are in order
against the common conflation of industrial laboratories with
applied science and with the development of new products (and
the equation of any of these with economic success) (Edgerton,
1996b).

First, some of the early industrial laboratories, such as that established by Thomas Edison in 1870, might best be seen as factories for invention; science was marginal. Inventors had come to the fore because competition between telegraph companies had created hopes that technical improvements would secure relative advantage. Edison established a laboratory because he had gained a contract and recognised that he needed a lot of new devices, and fast. Particular industrial conditions then favoured rapid (synthetic) invention, but that was not the only specific contextual factor. American firms developed industrial laboratories *c*.1900 in part because of anti-trust legislation which made it difficult for them to buy up firms which held patents they wished to acquire (or block). In Germany, industrial laboratories had developed only after 1876, when changes in patent law prevented the customary 'copying' of new dyestuffs. Such features of the 'regulatory environment' are always strong determinants of technoscientific patterns.

Secondly, much industrial research and development continued to take place outside 'laboratories'. Inventors did not disappear, and analyses could be conducted and experiments mounted in factories or in the field. Indeed, as Fox and Guagnini have claimed, the generating station that Ferranti built at Deptford to supply power to London could reasonably be regarded as one large, complex experiment. It cost a fortune, the technologies were untested on that scale, and it took years to get into working order. Some experiments would not fit into laboratories! (That is one reason for analysing STM in terms of its *projects*, rather than giving primacy to its *places*.)

Thirdly, we need to note that laboratories were not necessarily for the creation of new products; most company laboratories were established to carry out analyses for the better control of production, and that remained the major function of most of them. They might be used to investigate modifications of processes or materials, and this task might approximate 'invention', but *very* few laboratories before 1900 focused on novelty. This is *not* a sign of 'backwardness' – most industrial advance was (and is?) by incremental improvements achieved 'on site' or through analytical/ testing laboratories.[3] And, of course, even where 'synthetic' work was central, analytical functions remained important, as we saw in the last chapter for experimentalist laboratories.

In sum, industrial laboratories were important for many forms of analysis, of product and of processes, and they became important in some industries for invention/synthesis and 'development'; but not all of these activities were to be found in any given laboratory, and for none of them was a laboratory a necessary condition. *And yet*, in some companies, the industrial laboratory became the site where all these activities could be brought together, where all the relevant ways of knowing and making could be mobilised and linked with universities and government agencies. The sum of those relations was industrial, synthetic technoscience.

Electrotechnics and industrial laboratories

As noted, the key American exemplar of systematised invention was Thomas Edison, a self-educated telegraph operator turned inventor, who knew little mathematics but was widely read in practical science. In 1870 he won a contract to improve the New York stock-ticker system, and to that end he set up a private laboratory with fifty employees. The laboratory, expanded in 1876 at Meno Park and in 1887 at West Orange, was his tool for systematic invention. By making a business of inventing, Edison managed to put together all the elements of an electrical supply system. By the late 1870s he was seen as the chief author of a technology set to transform the world; he was the hero of the 1881 exhibition in Paris which launched electrical lighting in Europe (Fox and Guagnini, 1998: 125).

His later facilities included extensive libraries (for information) and testing facilities (mostly for analysis), and at their heart was a laboratory for invention where new combinations could be tried. Edison 'invented by design' to fill 'gaps in the market', his methods were based on 'cut and try'; but by the early twentieth century, in his laboratories and in those of his imitators, professional training in science became more important – to exploit new physical discoveries, to keep better contacts with academics and to recruit talented staff (who were more likely to be graduates as the universities expanded).[4]

An excellent study of the General Electric Company laboratories founded at Schnectady in 1901 reveals the closeness of academic science to certain new industrial sites. For over forty years one of their key investigators was Irving Langmuir, who had graduated in practical science (metallurgical engineering) at Columbia and then done a PhD in Göttingen with Walter Nernst, studying the dissociation of various gases around a glowing platinum wire. Nernst was a world leader in thermodynamics, but also interested in practical problems; he devised a new type of electric lamp that proved very successful. At GEC, Langmuir worked on many key areas of physics, including surface phenomena; he gained a high reputation and a Nobel Prize for his 'science'; but to regard his industrial work as applied science is probably an 'academicist' distortion. It seems to better fit the image pursued in this book – the creation of 'model set-ups' that could be exploited for various purposes, *both* for principles and for practice.

Langmuir began at GEC by trying to increase the longevity of light bulbs, studying various kinds of filament and discovering that residual gases adhered to the glass of the evacuated bulb. The experimental set-up was much the same as Nernst had used for investigating gaseous equilibria around a hot wire. Langmuir was led to investigate the dissociation of hydrogen molecules into atoms; he was also led to propose that bulbs be filled with inert gases rather than evacuated – a method of huge commercial importance. Further studies on the dissociation of hydrogen led to the invention of the atomic hydrogen welding-torch (Reich, 1985; Süsskind, 1973).

One could tell similar stories for industrial research in interwar England, not least for Metro-Vickers in Manchester, then famed for its industrial scholarships and for laboratories in close relation with the best academic facilities. But first we turn to the other field of early and dense interaction – the dyestuffs-pharmaceuticals complex.

Dyestuffs and pharmaceuticals

The key, again, is synthesis – of new molecules, a few of which had commercial potential. The special conditions included German chemical companies searching for new lines of business, an abundant supply of chemical talent (from university courses which had been boosted for analysis – in pharmacy, medicine and agriculture, as well as industry) and, as we noted, new patent laws in the 1870s which protected discoveries of processes as well as of products. To these we might add the general conditions of the dyestuffs and pharmaceutical industries – the perceived need for new markets, and a premium on novelty arising in part from 'fashions'. Think of the vogues for new colours for dresses, or the enormous expenditure on patent medicines. Note also the discovery of bacteria from the 1870s and the chance discovery of immunisation – both opening new approaches to therapy for companies that had the scientific and medical contacts.

But let us not anticipate. When synthetic dyes were first discovered it was by accident, by W. H. Perkin, a protégé of Hofmann at the Royal School of Chemistry. His discovery became famous, but it did not lead directly to a new industry because there seemed to be few other such dyes to exploit, and dyers were already producing novelties by traditional craft methods. However, by the later 1870s, several new classes of dyestuffs had been discovered and several German companies were collaborating with university chemists as consultants and to supply graduates. The companies had begun to send graduate chemists to work in university laboratories when they wanted to develop new products. The chemists 'at the works' were then largely engaged in analytical work – on the firms' own products and processes, and on other firms' products that they might wish to copy, or at least to understand. But the new patent laws placed a premium on 'home-grown' products, and the recent discovery of azo-dyes opened new pathways. Some firms 'pulled back' their employees from collaborating universities, and concentrated them in industrial laboratories designed to develop new products – backed by analytical and informational services. Invention here, as in Edison's facility, became 'mass-work'. Lots of chemists tested lots of compounds – few succeeded – but in a market built on fashion, one good novelty was worth a lot of money (Beer, 1959; Fox and Guagnini, 1998: ch. 1; Travis, 1992).[5]

By the end of the century the pattern was well established and reacting back on the educational system. Technical directors of major firms pushed for more practical courses in universities and colleges. When they failed, they supported new ventures in the new century, including a chemical institute under the auspices of the Kaiser Wilhelm Society, an umbrella for institutions meant to boost German industry by encouraging closer relations between principled and practical aspects of research and teaching – such as seemed to be developing in the USA.

The methods for dyestuffs were also used for new pharmaceuticals (or new versions of old remedies), which included over-the-counter remedies such as aspirin, most of which were palliatives to diminish symptoms. However, as industrial chemical laboratories were establishing themselves, medical science was undergoing something of a revolution – bacteria were being isolated for most of the common epidemic diseases –a new form of medical analysis that isolated *causal* elements, rather than tissues or cells that were affected by disease. The causes could be seen under microscopes, at least when dyestuffs were used to colour them (Lenoir, 1988).

Here was a nice connection. Dyestuffs attached themselves to fibres from the cotton plant and/or hairs from sheep – that was their *raison d'être* – so, they could also be used by investigators looking through microscopes at the tissues and cells of animals and plants. Natural dyes such as haematoxylin were part of the tool kit of the microscopists. Bacteria were simple plants, studied by microscopists; they, too, could be stained, indeed their staining properties helped in their classification. Just why particular stains attached to some biological materials and not others was unclear, but the fact was evident, and in the age of bacteria it attained fresh significance. Could one find a dye or other chemical that would attach to bacteria and kill them, without killing the cells of the animal or person who hosted the bacteria? Hence the search for the 'magic bullet' as envisaged by Paul Ehrlich, whose story is worth telling – for in such lives the elements could mingle without characterisation as either 'science' or 'technology'; and by such lives the technoscientific networks were created.

Remedies for/from microbes

From his student days in the mid-1870s, Ehrlich was fascinated by microscopy, especially the use of new dyes as stains for microscopic specimens. As a medical student he pursued these interests, surrounded by several key creators of the new 'bacteriology'. As a clinician he continued research on the selective affinities of dyestuffs, blood cells and bacteria. One of his dyestuffs, methylene blue, stained nerve fibres, and he showed it could also act as a painkiller. When in 1891 he discovered a stain for the newly discovered parasites in malaria, he administered the dye to two malarial patients, with some success. After a marriage to a textile heiress (note the synergy) and a bout of tuberculosis (note the 'incentive system'), he set up a private laboratory to investigate immunity – another kind of specificity, first uncovered by Pasteur's work on vaccines.

Pasteur had found by chance that weakened bacteria could induce resistance to disease (as cowpox inoculation had been known, since 1800, to produce resistance to smallpox). Pasteur's Institute soon showed that the immunity was in the blood – that proteins (antibodies) were produced which attached to the bacteria (antigens). Here was another new form of medical analysis which could be pursued in laboratory animals, and which had obvious therapeutic potential. It was Ehrlich who showed that various toxic proteins could also act as antigens and stimulate the production of anti-toxins. His studies, in turn, were connected with those of Robert Koch's associates, who pioneered the use of anti-toxins as treatment for diphtheria – a very nasty disease occurring mostly in babies and small children. Although not without its critics, then and since, diphtheria anti-toxin has usually been counted as one of the key therapeutic fruits of the new bacteriology (Weindling, 1992).

In 1895 Ehrlich was approached to direct a 'serum station' tasked with investigating and *standardising* anti-toxic sera. Standardisation was a problem, in part because the sera could not be analysed chemically for effectiveness; they had to be assayed in tests on animals. One key role of governments in the new 'biological' pharmaceuticals (vaccines and anti-toxins) was to help standardise them, so doctors knew how much they were giving. As we shall see, that state function has been recurrent in twentieth-century technosciences, and it has often provided a basis for adding some experimental work. Indeed, in 1899 Ehrlich was given a new (state) institution in Frankfurt, where he completed his diphtheria studies and began many new researches on chemotherapy. In Frankfurt he was close to a dyestuffs company, Farbwerke Cassella and Co., with which he had already collaborated. They made compounds to his specifications; he discovered that Trypan Red stained the trypanosome parasite and appeared to cure infected mice. The most famous of his products was Salvarsan, hailed as a cure for another parasite disease – syphilis. It was dangerous and difficult to use, but it seemed to herald a new world of rational therapeutics (Dolman, 1971; Lenoir, 1988).

We note the elements which were combined in Ehrlich's work – academic research, clinical research and practice, state support for new remedies and standardisation, and collaboration with chemical companies. This was the recipe for the new pharmaceuticals, taken up in the USA, and then in Britain and France, chiefly after the First World War (Davenport-Hines and Slinn, 1992; Galambos and Sturchio, 1997; Goodman, 2000; Liebenau, 1987; Weatherall, 1990). By then, the British government had established the Medical Research Council, which promoted research, standardisation and clinical trials (Austoker and Bryder, 1989). At this time, industrial laboratories were characteristic of most leading 'technical' firms in the 'new' chemical, pharmaceutical and electrical industries, even in Britain

(Edgerton, 1996b; Edgerton and Horrocks, 1994; Sanderson, 1972).

Science and industry in and after the First World War

It is perhaps in warfare that we can see the most striking extensions of technoscientific networks in the service of governments. The best known example is the mobilisation of scientists in North America to produce the atom bomb during the Second World War, but first I want to retail a brief case study of the most controversial aspect of science in the First World War, the poison gas programme. I do so partly to illustrate the scale and complexity of the mobilisation, and partly to link with the institutions and themes that we have discussed for industrial technoscience at the end of the nineteenth century. The First World War was the great industrial war (Pick, 1993).

The first use of poison gas was by the Germans in April 1915, when a huge cloud of chlorine was released from 6000 cylinders, to drift across the trenches of the French-Algerian forces. From the beginning of the war, both sides had been considering the use of irritating 'tear' gases; the German government took advice from the leaders of its scientific and industrial communities, including Walter Nernst, whom we mentioned above. Research into the various possibilities was conducted at the Kaiser Wilhelm Institute for Physical Chemistry and Electro-chemistry in Berlin, one of the turn of the century institutes developed by the state in partnership with industry to perform 'useful' research which seemed beyond the capacity or the inclinations of university departments. The chief author of the poison gas programme was Fritz Haber, a well-known chemical technologist, the head of another Kaiser Wilhelm Institute which became part of the military command for the duration of the war (Mendelsohn, 1997).

The USA and the UK also organised investigation and production of poison gases. The USA used its Bureau of Mines to establish a chemical warfare service based at the American

University in Washington, DC. One of the leaders in the work on mustard gas was J. B. Conant, later a major figure in the atom project (and a sponsor of post-Second World War research and teaching in history of science). The production plant for these deadly chemicals was not in Washington, but in the industrial city of Cleveland, Ohio. In Britain, many university laboratories were commandeered; and providing remedies against the gases became a major research programme for the newly established MRC. Here it could draw on previous research in respiratory physiology, performed in connection with the coal-mining companies (Sturdy, 1992a).

As Everett Mendelsohn underlines in his account of poison-gas programmes, few if any of the scientists had any qualms about the morality of chemical warfare. Conant was clear that poisoning soldiers with gases was morally indistinguishable from blowing them apart with explosives. The majority of leading German scientists saw German militarism as the spearhead of Western civilisation, and the allied scientists mobilised against the 'threat of the Hun'. The internationalism which had been increasingly prominent in STM since the 1880s crumbled rapidly; the power of nationalistic, state-directed technoscience was becoming horribly apparent.

In the First World War, much more than in the Second World War, most of the technoscientific programmes were improvised by government and military agents whose prior experience of technical programmes was limited. Famously, one of Ernest Rutherford's brightest assistants in the Manchester laboratory went to his death as an ordinary soldier in the trenches. But the scale of the mobilisation became huge and the British state gradually learned to use its STM talent to better effect. It stopped sending physicists and chemists to fight in the trenches, and used them instead to improve ballistics or to work out techniques for gas warfare.

The medical services, too, came to be highly organised, giving scope to new would-be specialists in cardiology or neurology. A hierarchy of casualty stations and hospitals was established, reaching from the trenches to the cities back home, where hundreds of public buildings and private homes were taken over as hospitals. At least in principle, this was the very model of medicine as *analysis and rational production*. In former tuberculosis sanatoria and mental asylums, special war hospitals were established in which doctors worked out new kinds of medicine – 'functional' approaches, based on experimental physiology and intended to get men back to the front (Cooter, 1993a; 1993b; Sturdy, 1992a). Though most of this huge machine evaporated at the end of the war, the research institutions emerged stronger, and some doctors had acquired a taste for 'organisation' that was renewed during the next war in debates about a National Health Service.

The effects of the First World War on British industry were rather similar. When it could no longer import German fine chemicals, pharmaceuticals or optical glass, the British government had to set up research teams and factories to manufacture substitutes – hammering home the lesson about German technical superiority which had been a regular complaint of British scientists since the 1880s. Huge supply networks were established, and researchers were mobilised to improve the workers as well as the products. The munitions factories, where most of the employees were women, served as laboratories for the study of fatigue – physiological medicine on the shop floor (Sturdy, 2000). When the war was over, some of the new organisation remained; the government set up a research council to fund scientific research in universities, and it supported research associations and laboratories for the major industries. In a sense, British industry was spurred to 'catch up' with German patterns, and perhaps especially with 'American' firms such as Westinghouse, Ford and Burroughs-Wellcome which had become major players in Britain in the years before the war (Alter, 1987; Edgerton, 1996b; Edgerton and Horrocks, 1994).

Historians have paid insufficient attention to the *ending of wars*, to what remained of the wartime structures, to the differences between peace-time institutions 'before and after'. The long-term effects of the First World War on technoscientific networks differed between countries, but generally they were less than for the war that followed. In Britain and France, the connections made in the First World War helped create a series of state-funded research institutes, for civilian and military programmes. In the USA, most of the military-STM links were abandoned, as part of the distancing from a dreadful conflict. In the 1920s the most conspicuous funders of American public research were the foundations (especially the Rockefeller Foundation and the Carnegie Foundation[6]), rather than the Federal government; the usual vehicle was university departments, which became increasingly oriented to research rather than teaching (Geiger, 1992; 1997; Mendelsohn, 1997). By the 1930s, in most Western nations, there were extensive and intensive bonds between *universities and companies* in chemistry, electrical technology (including electronics) and in the new pharmaceuticals.

In each country, in each such field, a handful of key researchers and consultants moved easily between universities, companies and government agencies. Henry Dale in the UK and A. N. Richards in the USA were 'everywhere' in pharmaceuticals (Swann, 1988). The Cavendish physics laboratory in Cambridge enjoyed close relations with Metro-Vickers's laboratory in Manchester (Crowther, 1974); the company supplied equipment and sought advice, Cambridge supplied prestige as well as information. The laboratory of Imperial Chemical Industries in Cheshire sometimes seemed like an Oxbridge college, and had contacts to match. (By contrast, their dyestuffs laboratory in north Manchester seemed lower-middle class and more obviously 'provincial' in its connections – Reader, 1970, 1975.) And if we look for technoscientific networks in inter-war medical services (rather than the pharmaceutical industry), the clearest example would seem to be radiotherapy for cancer – a small field, but changing rapidly in ways that contrasted strongly with the 'liberal' patterns of most of British medicine. It will serve rather nicely to illustrate the possibilities of technomedicine, even before the Second World War.

Radium, when introduced around 1900 had been the epitome of 'free market' medicine; it was seen as a 'general tonic' and incorporated into patent medicines, albeit in very small quantities. Larger quantities were bought by medical charities and by public subscription in the hope of 'burning' cancers, but radium was very dangerous and very expensive. From about 1910, central government was increasing its support both for scientific research and for public hospitals. By the 1930s, control of most of the medical radium was vested in the government and the major radiotherapy centres were regulated by the National Radium Commission. The centres were controlled by a new sub-profession of medicine – the radiotherapists – and linked both with physics departments in universities and with companies supplying X-ray generators etc.; they employed 'medical physicists' to help control dose regimens and radiation hazard. The Manchester centre was supported by municipalities across the region because cancer was then emerging as a 'public health' issue. Cancer had already been a field for organised biomedical research since *c*.1900, often with charity funding (Löwy, 1997); the Manchester research laboratory was attached to the radiotherapy hospital, and was both experimental and analytical. This complex was peculiar in the scale of its work, but it was widely admired as a model of 'scientific organisation' (Pickstone, 1985; Pinell, 2000).

We may conclude that, by 1939, if you had wanted to compare *'new industries'* across the different nations, then you needed to include these academic-industrial and medical complexes. Had you wanted to compare *academic provision*, then you needed to take account of governmental support for 'useful' research (which increased rapidly between the wars), as well as industrial support in money and equipment. In your survey of such linkages, you would have noted that *instrument-makers* served both universities and industry, and that *standardisation institutions* were crucial to these technoscientific complexes. So too was empire. In Britain and France, imperial engineering, agriculture and medicine, both governmental and commercial, employed a high proportion of the country's STM graduates, and provided a major field for university researchers (Farley, 1991; Worboys, 1976).

Yet that was not the whole picture. The techno-scientific linkages were often opposed; many academics and doctors resisted commercial 'encroachment' and even state support. Scientists in industry were excluded from some scientific and professional societies, not least in the USA, and a career in industrial science often seemed 'second best' to university or government service (Dennis, 1987; Swann, 1988). In Britain, technical education was rather '*infra dig*'. Oxford had finally taken science seriously, usually by importing talent from provincial universities, and big new cartels, like Imperial Chemical Industries, were respectable sources for research support and for employment of graduates – but still, too close an interest in technical matters was not good for an undergraduate's prestige. At Oxford, as a student or a fellow in chemistry, you were well advised to play down the 'stinks' and play up your musical talents (Morrell, 1997a).

Leading English physicians were well advised to keep abreast of medical research, but to subordinate it to the claims of experience and clinical judgement – to the *individuality* of both doctor and patient that was characteristic of biographical medicine (Lawrence, 1985; 1994). Specialists seemed to threaten this order of medicine, especially if supported by salaries rather than private practice. They suggested a world where the state might impose a 'division of labour' – each patient divided according to his diseases; each doctor specialising in particular conditions. (Analysis and rationalised production triumphing over biographical medicine and the liberal professions, as in the radiotherapy centres.) In most Western countries, supporters of the liberal order of medicine responded by stressing holism, natural history and humanities (Lawrence and Weisz, 1998; Timmermann, 2000).

Technosciences in and after the Second World War

This dynamic between the organisations of technoscience and the older, more independent forms of industry, medicine and university research was further transformed by the Second World War and its aftermath. The medium-term effects on the sciences were dramatic, especially in Britain and North America. The atom bomb is the obvious exemplar of a massive new project with major consequences for the post-war world; it required a mobilisation of scientists and engineers that dwarfed the previous scale of *academic* enterprises.

Since before the First World War, physicists had guessed that nuclear reactions could produce explosions, and by the late 1930s some researchers were checking the possibilities on paper. Physicists in Britain and America, fearing that the Germans might be developing an atomic bomb, pressed for a research and development programme, and serious efforts began in the USA in June 1941. They involved huge new facilities on several sites, some of which depended on the proximity of major hydroelectric projects. They also involved massive relocations of technical staff, including immigrants from Europe, and the concentration of state-of-the-art equipment from many of the country's best physics laboratories. Some of the management was provided by universities, most by the US army, including its engineering and explosive divisions. In technical terms, the project was remarkably successful; the morality of the use of the atom bomb on Japan continues to be debated (Mendelsohn, 1997).

It seems reasonable to think of the bomb project as the origin of the third great technoscientific complex of the later twentieth century – *nuclear science and technology*. In the USA it included the commercial nuclear companies, the massive facilities for atomic physics used by physicists from universities, the new programmes in medical physics and radiotherapy, and the geneticists' surveys of radiation damage in Japan (Owens, 1997). In the UK, it centred on the publicly owned Atomic Energy Authority, with major installations around Oxford and in the north-west – openly developing civilian nuclear energy and radioisotopes, and covertly making plutonium for bombs. In these countries, and in the USSR, the nuclear complex was a major part of a general reconstitution of science and industry after the Second World War (Gowing, 1964; Kevles, 1987; Rhodes, 1986).

Many other aspects of science and industry were also reformed, chiefly by governments. The US military became by far the largest sponsor of research in American universities; military projects were deliberately 'spilled over' into peacetime initiatives, and they served as examples for future national campaigns – to reach the moon or to cure cancer (Studer and Chubin, 1980). University-industry-government complexes thrived in aviation, ballistics and radioastronomy, and then in space exploration (Geiger, 1992). In several countries, state support was crucial for the development of computing, for even in peacetime the military was the major sponsor and user of large-scale data processing. In Britain, wartime planning and science was carried forward in newly nationalised industries and services, including the National Coal Board and the National Health Service (Gowing, 1964; Price, 1976). The universities, and the high schools which fed them, benefited from greatly increased state support, especially for STM. Public scholarships were now available to all eighteen-year-olds with the grades requisite for university entrance; meritocracy was linked to faith in education, especially in science.

In Britain and especially in the USA, wartime production of penicillin led to the rapid development of new antibiotics, and of research-oriented pharmaceutical companies; from them came new drugs acting on the cardiovascular and nervous systems (Galambos and Sturchio, 1997; Goodman, 2000; Parascondola, 1980).[7] The British National Health Service became a major funder of certain kinds of medical research, including salaried clinical professors in place of the part-time teachers who had relied on private practice. Clinical research became common in all teaching hospitals, often in association with pharmaceutical companies, and sometimes with the companies manufacturing medical equipment (Blume, 1992; 2000; Lawrence, 1997). In the USA, as we have seen, biomedical sciences benefited from the general increase in public funding for university research, and from the special research programmes instituted during the Korean war. All these developments strengthened the links between government agencies, pharmaceutical companies, universities and teaching hospitals.

Generally, in the UK as in the USA, the major STM programmes between 1945 and the 1970s could be described as *post-war* in a sense that is much more than merely chronological. The major technoscientific complexes were formed or reshaped in the Second World War (and by the cold war and the Korean war), and in all these post-war technosciences, the state was central. That is obvious for Britain; it becomes obvious for the USA when the military investments are included.

Coda

In our history so far, the post-Second World War decades are easily seen as the product of major social forces put in play by industrialisation and strengthened by international competition and conflict from the end of the nineteenth century. That is how the post-war decades presented themselves – their 'modernity' was compounded from scientific principles, rationalised production, functional aesthetics, medical prowess, professional expertise and the welfare state. To defend the West, some of these capacities had to be devoted to the cold war, not least to nuclear deterrence – a clean, if risky, form of power. All the technoscientific networks were crucial to this form of civilisation and to this self-image – for providing electrical power systems and communications, for new pharmaceuticals and the conquest of disease, and for nuclear energy and the balance of power. All these complexes were built from the employment of scientists by companies and state agencies, and by linkages with universities. Partly by the use of new kinds of electronic computing, these networks commanded huge amounts of information and considerable analytical capacities, as well as major facilities for systematic invention and experiment.

We should not forget (as enthusiasts for 'white-hot' technology often did) that many companies succeeded more by finding new markets or cheaper processes than by technoscience, or that the industrial health of nations might depend more on competent labour management than on industrial research. We should remember (as exponents of social medicine then insisted) that public health depends more on healthy diets, good habits and sanitation, than on miracle cures. And we have Vietnam to remind us that military success could depend on public morale rather than technical might. Indeed, it was partly such critical reactions which threw into relief the power of the 'military industrial complex' and its medical analogues.

Throughout the later twentieth century, the technoscientific complexes continued to grow, but from about the 1970s the politics of state and science changed considerably, perhaps especially in Britain. As we shall see in the next, and final, chapter, state support and direction declined *relative* to that of industry. Though the nuclear complex diminished by the end of the century, the electrical-electronic complex expanded around computers, and the pharmaceutical complex around molecular biology – but in both fields the directions came increasingly from commerce.

In Britain, by the end of the twentieth century, many public utilities had been denationalised and some state research agencies sold off, especially in agriculture. 'Business values' and professional management were being promoted in what remained of the public sector, including the health service and the universities. The government-funded research councils were being encouraged to think of themselves as part of the infrastructure of the national economy, and universities to focus on intellectual capital and the exploitation of discoveries. The rapid 'globalisation' of finance and business meant that a few international conglomerates dominated technological development worldwide. For these reasons, the end-of-the-century waves of technoscience – the informatics and the genetic engineering – were shaped by market values even more than would have been the case in the 1970s.[8]

It is in this context that 'public understanding of science' assumes a new importance. To those issues we turn in the next and final chapter.

Notes

1 On 'Big Science', which in my terms is mostly academic-led technoscience, see Capshew and Radar (1992); Galison and Hevly (1992); Krige (1997).

2 For early 'electrical technologies', see Heilbron (1979).
3 These points are also stressed by Fox (1998) and Guagnini (1999).
4 On Edison, see Hughes (1983; 1989); Josephson (1959).
5 Some useful essays are collected in Edgerton (1996a).
6 Much of their investment went to biomedicine, e.g. Rockefeller support for public health programmes and for early molecular biology, and Carnegie support for embryology.
7 For a case study of the reaction to cortisone see Cantor (1992).
8 On genetics, see Yoxen (1983); for public–private relations in pharmaceuticals see Walsh (1998).

8

Technoscience and public understandings: the British case *c*.2000

IN THIS CHAPTER, I LOOK back on the histories I have presented, and I look around at the condition of STM. I present a view from Britain *c*.2000, partly because it is difficult to get a more general view, but also because of the interest of the exercise. Other countries will look different and that will influence the way that readers elsewhere use this book. Again I use my local examples partly to emphasise the fact that we all live (partly) in localities, and that general histories can help us 'place' those localities. I use the now common notion of 'public understanding of science' to ask questions about the present social meanings of STM, and I develop some suggestions as to how 'ways of knowing' might contribute – both to the analysis of problems and to the construction of plausible remedies.

In the twentieth century as presented here, STM became increasingly dominated by technoscience – by the industrial-academic-governmental networks that manufactured 'knowledge-based' commodities. As we noted, the technoscience of the post-Second World War decades was in large measure a creation of national governments, especially the military. Since the 1970s, governmental influence has decreased (not least in the former Soviet Union) and commercial interests have become more concentrated and more global. Though the nuclear project may have declined somewhat, the electrical-informatics complex has grown apace, and so have the major pharmaceutical companies. Both complexes are offering to change our lives. The decades around 2000 are being presented as *revolutionary* – like the Industrial Revolution around 1800. Science seems full of transformative potential, and in that perspective *consumer resistance* now seems a major problem.

Here we approach what several historians have seen as the paradox of twentieth-century STM, especially medicine. Our understanding of the world and of our bodies is immeasurably deeper than in 1900; we have many techniques and remedies which were then only dreamed of, and many which were then inconceivable. And yet science and high-tech medicine are mistrusted, maybe more than in 1900, certainly more than in 1950 (Porter, 1997).

As we all know, this 'paradox' has many components. Western publics are much less obedient to authority than in 1950 – that is evident in matters of law, interpersonal behaviour, religion, etc., as well as for science. In 'matters of opinion', if not financially, British society is more egalitarian and less hierarchical. The media, whose 'discretion' could be relied upon in 1950 by governments and professional bodies, now place a premium on 'exposures' and investigative journalism; for better or worse, newspapers can sell copy by presenting genetically modified plants as a danger to health. From the 1960s, 'minority groups' that had been marginalised or subordinated have campaigned for their rights; the women's movement, gay rights, and campaigns for the new 'ethnic' communities have made Britain much more pluralist than it was. In these respects Britain reflects much of the West, and indeed has followed patterns first seen in the USA.

In social politics *and* in STM, the twentieth century was in large part *the American century*, and there especially social politics were linked with questions about the authority of STM. Women objected to being but 'mothers' in a medical system still oriented to the values of (male) production and (female) reproduction that were central to state medicine through the first half of the twentieth century. Homosexuals rebelled against their medicalisation, and blacks against the hierarchy of race which medicine had reflected and legitimated from the nineteenth century. For 1960s radicals, the 'consensus' politics of the 1950s and the high regard for professionals, especially scientists and doctors, suggested conformity – even social control.

Some writings on medical history came directly from this ideological realignment, notably the work of Foucault in France, and the associated studies prompted by the anti-psychiatry movement in the UK and the USA. Writings on 'ecology', notably Rachel Carson's (1962) *Silent Spring*, helped create a new movement for the defence of the environment (Bramwell, 1989; Sheal, 1976; Worster, 1994). Romantic attitudes to nature were resurgent and combined easily with a popular 'orientalism' in the promotion of alternative medical systems.

Generally, as in many such restructurings of popular opinion, new perspectives from the young and from 'outsiders' interacted with historical contingencies. Accidents in nuclear plants, which in the 1950s had seemed but technical hitches, now assumed a higher profile, partly because of campaigns against nuclear weapons. The thalidomide tragedy, when journalists discovered that babies had been born limbless as a result of medicines taken during pregnancy, shook public confidence and forced major changes in the regulation of drugs (and a major escalation in the complexity and costs of clinical trials). By the late 1960s, the two bright emblems of post-war STM – nuclear energy and wonder drugs – were casting long shadows. In the USA, the movement against the Vietnam war mobilised students and young academics, calling into question the military-industrial-academic complex which had been so important for the funding of STM.

This counter-cultural movement profoundly changed the social politics of the West, undermining the prestige of orthodox STM. But even in France, where student movements coincided with labour unrest and shook the state, the institutions of government (and of commerce) were not radically changed. The post-war consensus was, however, further weakened by the economic recession of the 1970s, especially the oil crisis of 1974 which revealed the vulnerability of Western economies to challenges from the Middle East and from the rapidly growing economies of the Far East. The political beneficiaries were the neo-liberal right, rather than the Romantic left.

In Britain, the radical critique of professional and governmental expertise was taken up by proponents of the free-market – an ideology which had been worked underground for thirty years; Mrs Thatcher and Mr Reagan wanted to roll back the state, and they succeeded; they looked to 'business values' rather than the traditions of public service, to accountants rather than civil servants. Mrs Thatcher reasserted Victorian values – of independence, philanthropy and the family – and denied the existence of 'society'. The move to the right in Western politics from c.1980 led to 'privatisation' of many state enterprises, including research laboratories, and the 'franchising' of governmental functions to 'businesslike' organisations. It greatly encouraged the growth of 'business management' as an ethos and practice; American management consultants were evident around British government from the early 1970s. Their kind of formalised instrumentalism – the mission statements, targets, audits, etc. – had become central by the 1990s, pushed into what remained of the public sector, usually against the wishes of state-employed professionals such as doctors and academics.

I recall the long faces of colleagues accustomed to advise governments on 'science policy' when they returned from a meeting in London in the mid-1980s: government ministers had declared that Britain had an 'information mountain', like the notorious 'mountain' of butter then surplus to the needs of the European Economic Community. To the usual claim that research was essential for future industrial success, Thatcherites had replied that Britain's problem lay in knowledge application rather than knowledge generation. At that time, too, universities were an easy target for expenditure cuts, both because of the recent history of 'unrest' and because they faced a demographic decline in the numbers of eighteen-year-olds.

This double threat to publicly funded 'science' from both sceptical publics and neo-liberal governments provoked some scientists to mount campaigns to improve the 'public understanding of science'. Their collaboration with sympathetic journalists, politicians and scientifically educated industrialists is perhaps best understood as a 'social movement', chiefly concerned to secure state funding for scientific researchers and to protect that enterprise from adverse public criticism. But their concern with the support of science has linked easily with the promotion of new technologies and with demands for improvements in scientific and technical education. It offers a useful point of entry for analysing present public meanings of STM and relating them to the concerns of the book.

'No one understands us'

Worries about the place of science in popular culture are hardly new. In the 1830s, young 'analysts' pointed to a British 'decline' as part of their campaign to gain the state support for science then evident in France and Germany. In the final third of the nineteenth century, a variety of enquiries, often prompted by proponents of science, argued for strengthening scientific and technical education at various social levels. Around 1900, 'science' was central to campaigns for national efficiency and the journal *Nature* was campaigning to raise the profile of science among the governing classes. After the First World War, while some social groups worried that science and technology were advancing out of control, others, especially on the left, came to see the rational application of science as the chief means to cure social ills and increase welfare at home and abroad. Debates continued after the Second World War – usually covering much the same ground as earlier debates (but without much sense of historical continuity). Problems were presented as newly discovered yet deep-seated; but one debate, from the 1950s and early 1960s still resonates (Alter, 1987; Edgerton, 1996b).

When the scientist-turned-novelist-turned politician, C. P. Snow, complained that the British ruling classes remained woefully ignorant of science, he was set upon by F. R. Leavis, the puritanical leader of a school of literary criticism. Leavis looked to literature, especially Victorian novels, as exploring and (thus exemplifying) the moral qualities which must remain central to public life. Like Matthew Arnold and many continental intellectuals, he saw sciences as 'instrumental,' as knowledge of *means* rather than *ends*, and thus as accessory rather than central to moral and political culture. This was common nineteenth-century formulation, as we saw in Chapter 2, but in post-Second World War Britain the reaction against C. P. Snow would seem to have been energised by the obvious fact that the majority of government investment in post-war universities and colleges went for science and technology rather than the arts. That preference has continued as central to educational rhetoric, and there is often embarrassment over the way that student preferences (and job markets) have chiefly benefited undergraduate education in arts and social sciences.

But if Snow intended to extend the political power of the sciences into a cultural hegemony, then he failed. The debate prompted a few initiatives in liberal education; several universities developed history of science or science-studies programmes in the 1960s and early 1970s explicitly to equip science and technology students for leadership (or political) roles. These programmes continue to be influential (and I acknowledge my own debt to the investments made in Manchester), but their best intellectual products were more critical than technocratic, part of the public scepticism about scientific and medical authority which mounted from the late 1960s.

To measure the failure of the technocratic ambitions of the scientific and engineering communities, one has only to compare them with the astounding rise since the 1970s of management sciences and various kinds of accountancy. It is *this* kind of technical training, offering generic, 'transferable' skills which has succeeded, not just in the worlds of business, but in public service institutions such as government departments, cultural agencies and universities. This *managerialism* has indeed established itself as the general education of the late twentieth century. Students in many disciplines are encouraged to do management courses, where in the 1960s they were encouraged to include 'liberal' options. The rise of management seems to be worldwide, and contests of managers with liberal professionals (e.g. doctors) or creative professionals (e.g. television producers) are central to contemporary redistributions of power. It is against this background that we must view the changing fortunes of 'public understanding of science'.

Science back in business

The public-understanding movement has continued to the present, but much has changed since the 1980s. The continuing 'globalisation' of business, the rapid development of the techno-scientific complexes around informatics and pharmaceuticals, and the projection of our age as a new industrial revolution have 'clarified' the roles expected of 'academic' science. The promotion of business methods across the public sector, including universities, has created a culture of 'output', of knowledge as commodity, which spreads the values of industrial science across the whole academy, even to the humanities. Whereas in the early 1980s scientists and doctors had felt deserted by government and in need of friends, now they know the roles in which they can expect to succeed. They are not to see themselves as advancing disciplinary knowledge, nor as representatives of a public interest, nor as educators bringing the best out of students; they are creators and managers of knowledge commodities.

More and more, the British government is now managing universities to reduce input and increase output, and the procedures are crude. The research councils which award public funds for research projects often see themselves as part of the infrastructure of national production; even social research has been managed primarily in support of economic development. In the 1990s, universities had to press hard to include community benefits as an alternative goal for research (and, indeed, for a university funding system focused heavily on student numbers and research publications). During the 1990s, as the medical charities emerged as a major force in British research funding and policy, they too became more concerned with commercial production, especially around genetic engineering. Via intellectual property rights, the charities expect financial returns on research investments.

But if these complex, rapid shifts over the last twenty years of the twentieth century 'clarified', for better or worse, the roles of the British academic scientist, they did not solve the problem of 'public understanding'. Indeed, as science has become more

closely identified with business on a global scale, so sections of
the public have become even more suspicious. Once they feared
that science was endangering them because of some internal
dynamic – some reckless pursuit of power over nature; today
they fear that scientists are in cahoots with big businesses –
messing up the world in pursuit of profit. So now we find worried
business people, industrial scientists, leading academics, charity
managers and government ministers of trade conferring and collaborating to 'sell' science. Public understanding of science has
become a corporate good and a corporate goal.

The continuing campaigns for science and the associated hype
around new industries, appear to be part of the reason for the
growing currency of popular science – as books and on radio and
television. Awards for scientists, journalists and playwrights, and
meetings sponsored to increase interactions between scientists
and writers have helped create a climate where some writers see
science as part of their engagement with a dynamic present, and
scientists see popular writing as a public service and a reputable
career strand – no longer a vulgar distraction. Scientific and
medical titbits are being introduced into news magazine programmes as matters of common interest, and the media can now
rely on a dozen or so high-profile scientists with well-honed pre-
sentational skills. Scientists aspiring to that status or worried
that they may be drawn defenceless into public controversies,
can learn those skills in dedicated training programmes. In such
a context, public interest in science, or public understanding of
science, is not to be read in terms of general 'levels', as if it were
the water in a bath fed by a 'trickling down' from scientists and
by natural 'upswellings' of public concern. Better to see public
concern as contested ground, where organised lobbies do battle,
and in which journalists of various kinds may grind axes as they
look for good 'angles'.

Surely such debate is to be encouraged, and it should be
enriched by commentators who can elucidate the issues involved
and weigh the arguments on each side. That is one responsibility of academics concerned with science studies, whatever their
personal attitudes and preferences. To that end, therefore, it is

important that funding for science studies not be dependent on the public relations lobbies for science, or indeed on anti-industrial lobbies. Part of the worry, of course, is the asymmetry here: environmental and public defence groups are paupers compared to the scientific industries and research charities. Moreover, as we have discussed, the latter interests are directly influential within government and in the state's higher education machinery, including those agencies which fund most of 'science studies'. We need to ensure that the public interest in disclosure and adequate analysis can continue to be satisfied by the scholarly community as well as by campaigning groups; that is part of the infrastructure required for a democratic politics in technical domains. But how shall we understand the ways in which 'publics' read 'meanings' in such fields?

The study of 'public understanding of science'

Academics who study in this area still operate by distinguishing an old view of the public from a 'new' one. The former was concerned with measuring how much 'science' was known by members of the public. Measurement was usually by interview or questionnaire; respondents were asked whether the sun went round the earth or vice versa. This view of the public is related to what the philosopher, Karl Popper, used to call 'the bucket theory of the mind' – minds were containers, so how much science, in the sense of 'scientific facts', did 'public' minds contain? By contrast, the (not so very) 'new' view regards minds as active rather than passive. All people, even laypeople, interrogate their environments on the basis of previous conclusions and assumptions – their hypotheses if you like. This approach to public understanding stresses the activity of the knowing person, and the way in which people become expert in that which directly concerns them. In this perspective, whether the earth goes round the sun or vice versa is not very important for members of the public – knowing the answer is little more than a trivial pursuit. But whether fumes from a chemical factory will damage your garden, or whether food additives will damage your baby – these are questions that matter, and which laypeople will pursue with vigour and acumen.

Note that so far in this section, public understanding of 'science' means public understandings of the world as interpreted by scientists. There is, of course, another set of meanings carried in the ambiguous slogan – it could mean the public understanding of 'science' as an activity of scientists, or of 'science' as a community of investigators.[1] How do the old and the new views of public minds relate to that ambiguity?

Under the old view, laypeople learn scientific facts more or less adequately from school, newspapers or television. They may also learn some facts about the methods of science, or about various scientific communities – such knowledge is on a par with the knowledge of nature, in that all is supposed *factual* and independent of the observer. Both kinds of knowledge are guaranteed by belief in the veracity of the reporting agency, i.e. 'science', whether the report concerns the physical world or that part of the social world that reliably reports on the physical. By contrast, if we assume that the public are enquirers, and see themselves as enquirers with a particular point of view, so they are likely to see 'scientists' as also operating within particular perspectives of their own. They may well see scientific findings as '*objective*' but as *less than the 'whole story'*. For example, even when patients' groups work closely with specialist doctors, they may still feel that the facts which patients need are best obtained from other patients, that doctors do not have the experience of dealing with a particular illness day in, day out – 'from the inside'.

Where discussion is polarised and opposing interests appear to be involved, then the 'scientific facts' become part of the dispute. Even if honesty and openness are evident, it will be understood that the production and presentation of facts is properly the object of scrutiny. Our antagonistic process then serves, hopefully, to maintain standards of evidence by facilitating expert scrutiny of claims by both sides. But for such debate to be fully effective it is also necessary that both sides are able to *create* evidence – by literature searches, interviews, analysis (and reanalysis) and by experiment. As Jerry Ravetz (1971) has stressed, the most worrying aspect of public scientific debate is not the suppression of evidence for a particular proposition, but the *absence* of such evidence when the research has never been carried out. In this respect democratic governments may be said to have a peculiar responsibility to fund research which may question powerful industrial and scientific interests, including the interests of government itself. This is the scientific correlate of the responsibility to facilitate debate by supporting the historical and social research community in ways that put its skills and knowledge at the service of otherwise disadvantaged groups.

In the remainder of this chapter I will try to show how the typology and schematic history presented in this book may be helpful to the various constituencies involved with such debates. Broadly speaking, I trace a path back over the concerns of the book – from technoscience back to meanings. First I discuss the roles of commerce and of public interest within technoscience, and especially in universities; then I use 'analysis' (and experiment) to reflect on the demarcation of 'scientific' expertise. I use 'natural history' to explore present and potential public roles of STM; and finally I return to questions of meaning and value.

The politics of technoscience

If the analysis of technoscience presented here is remotely accurate, then it would seem to follow that public understanding (and control?) of technoscience should be high on the political agendas of most nations. That conclusion seems to be borne out by the prominence in the media of many of the STM issues touched on here – the politics of health and health services, reproduction and ageing, food safety and environments, conservation and climate change, energy and armaments, etc. Adequate discussion of these issues and their connections would take a whole book, much longer than this one, but perhaps this is the place for some suggestions about modes of analysis.

If we think of technoscientific complexes as variously compounded of universities, companies and government agencies, then we can ask about the relative powers and prerogatives of the parties involved, and how they change. Since there seems little doubt that commercial interests are growing stronger rapidly – by direct power, through shifts of boundaries between private and public sectors, and by the private 'colonisation' of the public sector – the benefits and costs of that restructuring need to be examined. A sample of particular issues may illustrate the general problem.

One is the question of intellectual property in natural species, especially in the Third World. We have seen in Chapter 3 that exploration for commercial purposes was a major component of the growth of natural history and that imperial powers mounted expeditions and amassed collections for commercial reasons as well as for the demonstration of political and cultural power. The links between science and colonial exploitation are long established – but the recent developments pose new problems as well as old. In as much as species are being patented, they are appropriated as well as exploited; they are taken out of the public realm. Third World countries and their supporters argue for vesting the potential 'property' in the countries involved, at least for the purposes of preventing appropriation.

Similar issues have arisen over the patenting of sections of the human genome. Biological entities that most people presumed to be inalienably 'public' have been privatised. The extent of privatisation of the genome may have been diminished by the acceleration of public and charity investments in genomic analysis – thus limiting the amount which private companies are able to control – but one may wonder why the genome cannot be protected by adjustment or reinterpretation of law so as to exclude the patenting of natural entities. The boundaries of what can be patented have varied considerably over time and between countries; they could perhaps vary some more.

These issues have a history. In Britain after the Second World War there was much discomfort over the American patents on the preparation of penicillin – because it was a British discovery, made by doctors and scientists in the public sector, and because it was a naturally occurring substance urgently needed by millions (Bud, 1993; 1998; Macfarlane, 1979; 1984). In our present, part of the offence arises from the knowledge that the vast majority of the work that made possible the human genome project was funded by taxpayers and conducted by scientists in the public realm, who are now to pay for the use of discoveries.

Through patenting and the many forms of sponsorship, the tradition of public service research is now seriously endangered by commercialisation of university science. Perhaps our need now is less for *public understanding of science* than for a much wider and stronger *understanding of science for the public* – the roles that STM could and should play in the defence and development of public interests. In both the above cases and in others we have mentioned, the problem is to find mechanisms for the defence of the public realm against undue intrusion from commercial interests.

One answer is to refine and extend the regulatory and informative roles of government. In sectors where life was at stake, such as the safety of medicines, most advanced countries now have reasonably effective mechanisms. In some newer areas, such as the regulation of reproductive techniques, Britain has pioneered agencies which appear to work well in as much as they command public consent while keeping the issues at a distance from main-line party politics. It seems likely that public opinion will force the creation of more and stronger protection agencies, though they are usually resisted by producer interests.[2]

Such agencies typically involve expert professionals and laypeople, and they are open to scrutiny by the public, the media and campaigning groups. In as much as the Internet etc. can serve to make them (and their results) more accessible, and possibly to reduce the cost of duplication of enquiries across nations, they would seem to be a very promising part of the future of STM. How such protection can be extended to poorer countries, perhaps by United Nations action or by levies on major companies, remains unclear, except that we might try to think of global systems in which such protection would be as 'natural' as free trade is now claimed to be.

In as much as governments can be held to account for public protection against adverse effects of STM on individuals, we are likely to see extension of these regulatory roles, including agencies to ensure the probity of scientific publications now that huge sums may depend on getting the correct results or on disguising failures (Kevles, 1998). When more and more scientists are financially involved with commercial companies, the operation of peer review by research councils and charities may need public scrutiny, including public registers of interests such as are now obligatory for parliamentarians. One of the advantages of thinking across the range of STM may be to facilitate comparative study of the various forms of regulation – across sectors, functions and nations. Certainly, we need more – and more public – studies of this kind.

But governments are also involved in technoscientific complexes as purchasers. Here, too, Britain, following the Netherlands offers some encouraging examples. Agencies have now been created to assess new drugs and other medical technologies, not just for safety but for effectiveness and for cost benefit compared to others treatments. Presumably there is scope for much international flow of assessments of this kind, and much benefit to other potential purchasers, including individuals. Perhaps such agencies might be seen as continuous with the assessments (and consumer satisfaction data) collected by consumers' associations. Maybe the support of consumer information systems will come to be seen as a major governmental responsibility – part of the 'natural history library' required for effective living in a highly complex commercial society.

All such agencies allow better public and private choices; they would seem obvious candidates as potential commissioners of the public sector research which we argued above was essential for proper public defence. Charities such as Greenpeace or Friends of the Earth, or user groups in medicine, or local charities concerned with particular issues might also be appropriate. By widening the range of interests and groups that might command such research, we may balance the intense present pressure which equates the public interest with economic development. Moreover, similar arguments could be made about Third World interests. Whereas once British anthropology was supported in the hope of improving colonial rule, now our 'overseas research' is strongly focused on potential markets. We can surely afford a wider definition, even of 'British interests'.[3]

But in this discussion of commercial and public interests, where do universities fit – for their teaching, and especially for their research?

Understanding public science

In previous chapters, universities have often been mentioned but their governance and their legal forms have not been discussed. We have presented them as most people now see them, as independent agencies receiving money from taxes but traditionally at arm's length from government. Of course, like so many traditional views, this one needs to be understood historically and in cross-national context; and we need to recognise that universities are now in flux. Historically, the German model is crucial for two features – the ideology of research as a human good, and the productive interaction between the pursuit of knowledge and the education of scholars.

In Germany, universities have always been state institutions; the freedom of university staff to teach as they wished and of students to move between universities were key features of the nineteenth-century reforms – part of the ideology of research to which I have several times referred (Proctor, 1991). The usual image of the German system since 1800 (excepting the wars and the years of National Socialism) is of self-governing institutions competing for staff eminent in research – a system of competition and emulation which encouraged and supported talent and led to world eminence in most fields of research. Historians have questioned this picture, but present studies of nineteenth-century appointments tend to show that commercial and political 'interference' acted more to reinforce than to damage excellence. For

example, Turner has shown, as any academic might have sus-
pected, that left to themselves some universities and faculties
tried to appoint from their own staff and protégés, rather than
importing talent. In such cases, the Prussian Ministry of
Education protected the system against inbreeding (Turner,
1971). Other historians have shown that the pursuit of science
was not as disinterested as sometimes presumed; when states set
up a chair in chemistry, say, it was often with an eye to the
improvement of local industry or agriculture (or medicine). But
again, since it was usually assumed that prestigious chemists
would be most effective economically, there appears to have been
little conflict between commercial motivation and the appointment of staff with high reputations.[4]

The American system is a mix of educational charities (e.g. the
University of Chicago) and state institutions (e.g. the University
of Minnesota). From the later nineteenth century, the German
ideology of research was central to all the major institutions, but
it was combined with an American insistence on the practical.
The usual result was a broad education in principles, followed
by science-based professional training. American universities
developed rapidly in the early twentieth century when wars and
economic depression limited European advance, and after the
Second World War the elite American universities, and especially
the technical universities, benefited hugely from government
research investment. The system has been rich and diverse
enough to preserve a variety of ideologies, from corporate ally to
pure scholarship (Geiger, 1997), but the commercial challenge is
becoming acute – highlighted when a whole department within
the University of California signed a deal to become, in effect, the
research arm of a bioscience company.

In the UK, the oldest universities were religious corporations;
the nineteenth-century foundations were mostly secular charities
– self-governing entities, funded by donations and student fees,
and jealous of their independence (Sanderson, 1972). State
support for research only became important in the twentieth
century, especially after the Second World War. Through the
1950s and 1960s, universities expanded rapidly and research

was well funded, largely on its appeal to fellow academics. Undoubtedly this meant that universities became less responsive to their localities – in terms of industry, public services and culture. The effect was accentuated by the 'academicisation' of the major technical colleges which had been funded through local government, but which were transferred to the university sector, culminating in the 1980s when all the former 'polytechnics' were brought under the same funding council as the old universities. The consequent wide disparities within the sector is one of the reasons for the elaboration of formula-funding mechanisms, for the diminution of support per capita (levelling down), and for systems of micro-review and heavy bureaucratisation.

I have suggested that increase in direct management and the stress on commercial values can both be regarded as aspects of the incorporation of the universities into networks for the production of knowledge commodities. What then is at stake, in terms of the maintenance of public service and public knowledge, and how might these public interests be defended? We have noted above that part of the solution could be a more open and diverse system of research support, where public-defence agencies and charities, etc. might commission short-term research, and might join with academics to put forward proposals for research aimed at public goods other than increased production and consumption. The membership and range of the research councils and research charities might need extending.

Widening the range of potential research funders, and reinforcing inputs from wider publics, might also help meet the need for expertise and experts who are not compromised by commercial involvements. But perhaps we also need to rediscover the value of learning and of research as an end rather than a means to more production or, even, safer environments. When the then 'new universities' of industrial Britain were founded in the later nineteenth century, they were meant to contribute to industry, but they were also meant to offer and develop values other than utilitarianism. The motivation derived, in part, from the German tradition of research; it overlapped with the drive for civic improvement or art galleries, as well as for better technical colleges. In their STM, as well as in the arts and humanities, universities need to find ways of reclaiming that cultural heritage and those educational values. That may mean trusting professionals; it surely means encouraging the huge potential of universities for open-ended interdisciplinary dialogue, formal and informal. Universities as communities are being undermined in Britain because internal accounting and relentless 'reporting' occupy the space once allowed for creative conversation. Business management produces businesses.

A third general remedy is more experiment. To widen research demands, to recognise learning as an end, and to better serve local and global communities, universities and other public bodies need more room to be original. In the centres of most of the major cities of Britain, there are thousands of creative staff and tens of thousands of bright young people in institutions that are still largely funded by the public. If we want to be 'scientific' about developing regions and universities, let that 'science' include lots more experiment; 'scientific management' here should encourage institutional novelties, perhaps with rewards 'to follow'. Standardised systems of surveillance have the opposite effect.

In sum, it seems to me that the political problem now is not the *public understanding* of science, but rather the maintenance of *public science* – creating and using knowledge to best serve public interests. But we need to underline that this is an issue for *all kinds* of knowledge, not just for 'science' or STM. The limits of 'science', especially in the English language, are unhelpful. In this book, analysis and natural history clearly reach beyond the normal British boundaries of STM; perhaps we need to rethink those boundaries. We could probably benefit from thinking of STM more as we think of the social sciences and humanities (and vice versa).

Analysis and the bounds of 'science'

English-speaking countries use rather narrow definitions of science compared to countries that speak French or German.

'Science' in English is restricted in its subject matter – used alone it usually refers only to natural sciences, rather than to moral or social sciences; history and literary criticism are not normally called sciences at all. This 'width' restriction then has consequences for questions of 'depth' – or ways of knowing. It focuses our attention on the methods supposed to be most characteristic of natural sciences, i.e. laboratory manipulations, especially experimentation. Analysis of economic data, say, which would be 'scientific' in other languages, has an uncertain status in English, and the natural history of riverside meadows, at the level of recording species throughout the year, is unlikely to qualify as 'scientific'.

Both restrictions (and their interactions) seem unfortunate. As regards width, it seems discriminatory and divisive to routinely distinguish between two classes of systematic investigation on grounds of their subject matter. Critical analysis of economic and social phenomena is prima facie as useful and demanding as analysis of geological strata or ballistics. We need better analytical indicators of human well-being as much as we need better indicators of environmental change; we need informed debates about the limits of economics, as we need debates about analytical medicine. It is no accident that such forms of knowledge developed together, as we saw in Chapter 5. Recognition of that (historical) commonality could do much to overcome deep educational divisions that limit critical analysis of social phenomena, as they also limit critical attention to the operations and results of natural-scientific communities.

One reason for the British 'exclusion' of social sciences may be the undue emphasis placed on experiment rather than analysis as crucial to 'scientific method' – especially as popularised after the Second World War. By stressing analysis and the pluralism of science perhaps this book will make for more *inclusive* usages and attitudes. For example, it is customary for government ministers in debates about food safety to demand 'good science' as the basis for their decisions (or delay). Such calls are markedly less common in matters of social or educational policy, where the individual opinions of ministers or functionaries often appear as sufficient authority for action. More symmetry might help – issues of safety cannot be decided only by 'science', at limit because they involve judgements about the 'worth of a life'; social policy might benefit from rather more *analysis* – and *experiment* testing.

Indeed, social experimentation might be one of the beneficiaries of the 'informatics' revolution. I wondered at the start of this book whether new 'ways of knowing' might be added to my list, not just by creative critics, but through the evolution of STM. One possible 'new way' is *simulation*, which is already widespread in STM teaching 'laboratories'. Some of the computing power once used for US military calculations is now available for civilian use, and capable of modelling the traffic patterns of a major city in full detail.[5] There surely are possibilities here, especially for experiments in the redesign of institutions which anyway use programmes for many of their bureaucratic functions. Could we simulate a variety of systems for 'experimental universities' that would enhance public accountability and professional creativity, without a myriad of particular reporting mechanisms that partly overlap but are never substitutable?

Here and elsewhere, talk of 'the scientific method' tends to mislead. As Paul Feyerabend demonstrated, it is very difficult, perhaps impossible, to codify the rules which are followed in really creative science – that level of creativity usually means altering the rules (Feyerabend, 1975)! The real commitment is not to rules of method but to the exercise of *imagination and critical faculties* – and that applies as much in dramatic productions or the development of social services as in neurosciences – or it should. This book is built on the benefits of recognising the pluralism of STM and the continuities between different 'subjects' that can be recognised as 'ways of knowing'. This is *inter alia* a tool for breaking down the supposed boundaries of 'science' and the supposed exclusivity of 'scientific method'.

To that end, I now go on to suggest that including 'natural history' as 'science' may also be beneficial – not just for certain very necessary types of investigators (e.g. taxonomists) but for the health of scientific enterprises and of publics generally. To include natural history is to include a wider public *within* the debate – as actual and potential participants, rather than as receptacles for expert assessments. The more we can see the *variety* of STM and its articulation with everyday life, the less likely we are to be mystified by technicalities. And at the end of the chapter, we move on from natural history to 'meaning'.

Publics and natural histories

As we have suggested, natural history is often left out of 'science' and relegated as mere 'information'. I have argued for a wide view of natural history, and for placing this descriptive, classificatory way of knowing at the heart of the scientific enterprise – as a cultural achievement on which analytical and experimental modes are built, and one which remains a major way of dealing with our world. In as much as natural history can be more easily learned and practised than analysis or experimentation, and in as much as it is more continuous with ordinary language and experience, it is more accessible to untrained publics. Natural history is often recreational and sociable; it is 'phenomenally' rich. Questions of amenity and/or safety, e.g. around pollution, are often to be researched at the level of natural history, and citizens have a right to ask for questions to be addressed at that level. Here I want to point up several ways in which our focusing on natural history can guide our thinking about public understandings of science.

Access and recruitment to science

There appears to be little social research on why students choose science (or why others avoid it).[6] Much would seem to depend on public perceptions of 'rewarding careers'; these, of course, are part of social history, and could usefully be so explored. Let me illustrate by personal anecdotes. In the late 1950s many of us chose 'science' as the default option at school – if you were bright and *could* do science, you did so. Only if you were innumerate, or uncommonly passionate about history, say, did you choose to spend your final school years (and therefore your university career) doing languages or humanities. We knew about the Russian satellite 'Sputnik' that had so jolted the West, we had been visited by university professors who lectured us on Britain's need for scientists, and many of our parents had heard of someone who had 'done well' as an industrial chemist, or some such.

Of course, there were also more positive and subject-specific reasons. Some chose science because they wanted to do medicine. Others were fond of some aspect of science that they followed or practised 'out of school'. Maybe you were interested in debates about evolution, or went bird-watching, or you had a microscope, a telescope or a chemistry set, or you made radios or took photographs and developed them yourself. These were the 'hobby' roots of many scientific careers. For the most part, they qualify as natural history – collecting, describing, arranging, or crafting apparatus to produce 'special effects'. Sometimes these activities were solitary pursuits, sometimes a bond between friends, sometimes pursued with local societies – e.g. for astronomy or ornithology. Such interests undoubtedly shaped the choices and the inclinations of scientists, as Jeff Hughes showed for British physicists (Hughes, 1998).

I studied biology partly from an interest in botany encouraged by a chemistry master (and amateur botanist) and by my maternal grandfather. At the start of the twentieth century, my grandfather had attended classes and field trips at the Burnley Mechanics' Institute, taught by an autodidact 'biologist', Ernest Evans, who wrote books on local history and natural history and was also known as a scientific poultry-breeder (egg-science again!). My Grandfather's brother went on from 'the Mechanics' to the Imperial College at South Kensington, and spent his working life as an entomologist in the colonial civil service. They sprang from a tradition of Lancashire textile workers escaping from the cotton towns to the surrounding hills (and maybe reading Wordsworth and Ruskin). Perhaps we need to see science as part of family histories as well as daily life.

Nowadays, 'popular science' is largely about/via computers. Young prodigies can make their fortunes; adolescents come to university with a facility in informatics that embarrasses ageing teachers. On the biology side, students come with developed interests in ecology or in the wonders of genetic engineering. The former are more likely to have direct, out-of-school experience – and they are more likely to be disappointed by the strongly 'molecular' orientation of most university biology departments. If science and technology departments want to increase the quality of input, or want students with interests other than technical-business, they could do worse than attend to public participation in sciences at the level of craft and natural history. Archaeology offers a model here: in Britain it is not a school subject, but it recruits well at university level partly on the basis of local and regional 'digs' which school students can afford (and via the corresponding coverage on television, radio and in magazines).

In recent years, most public discussions of science education have concerned assessment procedures and entrepreneurship. There is room for much more exploration of the roles of science education for individual lives and collective action.

Information and utility

In Chapter 3, I stressed the similarities between the natural-historical orderings of life's riches and our present concerns with superabundant 'information'. The technology now exists that can bring into any home or office, stocks of information beyond what could be found in any of the major libraries. Natural history collections, museums more generally and art galleries are all accessible virtually; so are data on diseases, medicines and environments. The Internet allows endless unpredictable access to the huge quantity of data specimens that individuals and groups throughout the West have seen fit to display; it brings a new world to armchair explorers, a new richness to our virtual 'environments'.

Natural history, however, is not just a matter of information or a logical or pedagogical prelude to other sciences – it is part of living. If there is any currency left in the idea of a *capable* informed citizenship, then the various kinds of natural history still have much to offer – for private as for public purposes. Part of the pleasure of the countryside is our knowledge of its inhabitants, its forms and its development; part of our citizenship is our understanding of our surroundings. Here, too, we may point to medicine, and the importance of patients' understanding the natural history of disease and its causation – and, indeed, the natural history of health.

For all the technology in medicine now, it remains true that the rules of health are simple and long known, and that attention to these simple matters can outweigh curative medicine in improving morbidity and mortality rates. A varied diet and a varied regimen, clean air and water, sufficient sleep, work, companionship and exercise – these remain the keys to health. To follow them without compulsion requires an intelligent interest in the health and routines of oneself, family and friends. Here the practice of self-observation, a kind of natural history, can be useful; so, too, is an appreciation of individual and social development, and of human *variety*.

All such were part of 'biographical medicine' that we discussed in Chapter 3, and especially of the Hippocratic tradition in Western medicine. They have been variously preserved and revived in West and East, often in conjunction with a developed interest in natural history narrowly defined. In seventeenth-century Britain, Thomas Sydenham became known as an advocate of a kind of natural-historical empiricism, both in medicine and in the study of plants. In the inter-war decades of the twentieth century, one of the conspicuous 'health centres' in London was initiated by doctors in the pursuit of a 'human biology' – a natural history of human organisms in their natural and social environment (Williamson and Pearse, 1938). More recent examples abound – from the 'natural history of disease' advocated by proponents of 'social medicine' around the Second World War, to the ecological orientation of the 'new public health' movement in the 1980s or, indeed, the close present association between environmentalism, new lifestyles and 'alternative healing'.

In as much as these movements, from the nineteenth century, were often in tension with 'medical orthodoxy' they are evidence of the *variety* of ways of knowing that might be called scientific, evidence for the *pluralism* that is central to this book. It is too easy to oppose the 'rational medicine' of routines and large samples to an unscientific individualism; we may need reminding that these are not the only options. We hear more and more about 'evidenced-based medicine', relying on clinical trials, the expensive and expansive descendant of the 'numerical method' of post-Revolutionary Paris; but 'mass medicine' is not the only way to be scientific, nor are statistical trials the only relevant form of 'experimentation'. In Chapter 5 we discussed the way in which physiological medicine in nineteenth-century Germany could sometimes appear as the scientisation of biographical medicine – perhaps using quantitative measures to study the equilibria of individual bodies. Such approaches may sometimes be 'objectifying', but they might also be a welcome means of exploring individuality. Standard doses of drugs may be good for killing bacteria, but for medicines taken over long periods, including psychoactive drugs, we may need rather more individualised frames of reference – practices based on the long-term observation of the patient, rather than statistical averages. The popularity of individual 'therapies' may suggest a need for more individualised versions of orthodoxy – medicine for racing cars, perhaps, not 'fleet' saloons; or a high-tech, egalitarian version of the 'connoisseur physic' we discussed in Chapter 3?[7]

To be effective, any such individualised practice needs skilled, informed patients. We have already mentioned the possibilities of the Internet, especially where there are active 'patient groups'. The mutual-instruction societies once built around books and newspapers are now in virtual space, linking the users of web sites, connecting all who would know more about their diseases or their dahlias. We are already seeing a reshaping of medicine, with many patients better informed than their doctors. For some conditions, e.g. coeliac disease, so much information is available 'on line' that the management of the condition has been effectively demedicalised; doctors may be needed for the diagnosis, but then you can do better 'on your own'.

Maybe the twenty-first century will come to resemble the eighteenth in the proliferation of the 'descriptive sciences' and a relatively open flow of knowledge between professionals and others. We can be sure that 'control of information' will remain a key political question, even as the bounds of our extended natural history extend still further.

Defending publics

In his path-breaking work on *Scientific Knowledge and its Social Problems*, Jerry Ravetz (1971) anticipated many of the points spelled out in this chapter. We have noted the problem of asymmetry in scientific debate when the two sides differ hugely in their resources for research; he also alluded to another related asymmetry – the different *kinds* of STM commonly deployed in debates over the benefits/hazards of biomedical and/or environmental initiatives. Characteristically, the new products or procedures are the results of technoscience involving much esoteric knowledge and experimentation (or they are so presented). Typically, the *critics* rely less on experimental science than on certain kinds of analysis and on simple observation (or natural history). This is in part because of economic asymmetries between the contenders – patients' lobbies or environmentalists rarely have the funds for extensive or elaborate research. But it is also a question of differences in approach, rooted partly in stubborn facts about ways of knowing. Well-trained chemists can predict the products of chemical reactions with a high degree of precision; they may be able to predict the toxicity of a particular product known to interfere with particular metabolic pathways; but the quality of prediction declines as the systems under consideration become more complex. We may not know, even 'in general,' how certain toxic metals are likely to 'circulate' in the biomass of a pond, and the problem of prediction is compounded when the questions relate not to ponds 'in general', but to a singular pond – or any other singular habitat or organism, the *relevant* characteristics of which may be very difficult to specify.

Here, as so often, medicine provides us with useful reminders. Predicting the outcome of a 'condition' in a single patient is often difficult, especially if we are dealing with complex chronic disorders. In medicine, because the systems are complex and singular, there is always an element of 'wait and see', supposedly absent in physics. But before we conclude that medicine is different, we might remember the considerable literature by historians and sociologists who have studied experimental physics 'as it really happens'. They have focused on the role of tacit knowledge, on the difficulties of replicating experiments and on the skill of 'discarding' experiments that did not 'work properly' (Collins, 1992). As Bruno Latour has emphasised, one key difference between experimental science and more worldly activities (including politics) is that physicists can repeat a test as often as they like, discard some results and present an 'average' of the rest. Asked to ensure that an experiment will work properly, first time, in a *live* lecture demonstration, they experience the uncertainties characteristic of all 'singular' human activities (Latour, 1993).

Generally speaking then, the more we are concerned with complexity and singularity, the more we resort to natural history. That is true inside laboratories where the successful completion of experiments depends on huge amounts of informal knowledge about the best ingredients or the best conditions. It is doubly true outside laboratories, where ingredients and conditions are not so easily controlled, and where consequences are likely to extend beyond the immediate participants. The defence then against scientific hazard (or the measures of benefit) is likely to include careful observation of all likely parameters, over the long term. Side effects of medicines are still discovered 'in use', no matter how elaborate the laboratory testing. The only way to be sure about the environmental effects of a chemical plant is by careful *monitoring*. It follows that debates about hazard and benefit will often turn on 'natural history'.

If we can focus on natural history as the accessible heart of many public debates, then the prospects for democratic outcomes would seem better than those derived from other models of science–public relations. For as long as we focus only on analytical and experimental knowledges that are remote from experience, for so long may 'publics' be disadvantaged. If it is generally recognised that observational approaches are likely to be persuasive, one way or another, then the advocates of developments will perhaps give more respect to the complexities and challenges of observing complex, singular systems, and publics will be encouraged to demand such data.

Yet again, the model is medicine. Assume that you are rich and powerful. When you talk with your doctor you want him or her to explain, at the level of *your* history, the likely effects of a treatment (including the likely *range* of effects). You are not primarily concerned with explanations of the genetic code or of cell division, you want to know how confident they can be in predicting outcomes for you, with or without the treatment. The same argument can be applied for environmental questions, except that most of the environments in question may be regarded as presently healthy – in which case 'treatment' requires more justification than would inaction! But of course, that conclusion depends on how you 'read the world'. Here, then, we return to the issues raised in Chapter 2.

Public understandings and world-readings

By stressing world-readings or hermeneutics in this book I hope to open up the debate around STM by emphasising an obvious fact, too easily ignored: that our primary relationship to nature, as to each other, is one of *meaning* – of morals and aesthetics, or of religion (for some). Questions of meaning underpin all our relationships with the world, including our appreciation and use of other ways of knowing. I have tried to show something of the history of 'disenchantment' whereby certain aspects of nature were 'objectified' and considered instrumentally or as technical matters; but for cultures, as for individuals, such 'disenchantment' should be considered as always part of a dynamic balance. In as much as we are human, we are always investing meanings. Of course, we must often abstract such meanings to facilitate discussion and collective action of certain kinds, but we should never forget them.

That is perhaps most obvious when we consider patients in medicine, and especially when we think of ourselves as patients. Then we look for meanings in illness – why has this happened to us? Did we deserve it? But we also restructure our social relations around illness, classically (for the 1950s) by adopting the sick-role, as described for acute illness by the sociologist Talcott Parsons (see, e.g. Currer and Stacey, 1986; Stacey, 1988). We become dependent on doctors and our carers, by temporarily redefining ourselves as patients rather than as self-reliant.

The 'sick-role', however, is not the only way of being ill or disabled and, at the end of the twentieth century, chronic diseases and disabilities seem the more challenging site. Though it is easy to mock the terminological preoccupations and the political correctness evident in debates around 'persons with disability' etc., we can recognise the importance of definitions in establishing *shared meanings* for discussion of individual and social problems. If we describe a person as 'mentally deficient' we place them on a scale of measurement; if 'handicapped' we attribute to them a probable failure of performance, without considering how that handicap might disappear (or even become an advantage) were the world to be remodelled a little. The disability rights movement has been remarkably successful in advocating and securing change in our common frames of meaning (Cooter, 2000a). Like the campaigns for ethnic rights, women's rights and gay rights on which it builds, it reminds us that the world of early twentieth-century medicine – its scientific accounts of racial and gender deficits and of perversions – was science from a single point of view (Kevles, 1985). As crude versions of genetic determinism daily echo that old eugenics, we need to remember that genes are but parts of biological systems and that, additionally, the 'outcomes' are always understood within systems of meaning that are disputable.

Much the same arguments apply to environmental questions. To understand debates on these issues we must understand the systems of *values* that are in conflict, and how scientific findings may assume different meaning in different frameworks. If we believe that human happiness, generally speaking, is best advanced by scientific and industrial progress as currently conceived by the relevant powers, then our attitude to genetically modified crops will probably be positive, if 'cautious'. We will want mechanisms to guard against 'side effects' – our model is likely to be *medical* because recent Western societies have generally favoured pharmaceutical research provided that mechanisms were in place to guard against hazards, including those which could not have been predicted. But for many opponents of genetically modified crops, the real issue is *not* the *additional* hazard of the particular technology but the likelihood of its accentuating tendencies in agribusiness which are seen as harmful already – destruction of habitats, loss of genetic variety, standardisation of foods, Third World exploitation, the subordination of small farmers and peasants, etc. To ecology-activists, genetic engineering merely accentuates the polluting, standardising and controlling tendencies of big agribusiness. Against the threat to biological and social diversity, the prospect of 'cheaper food' in rich countries seems derisory.

One could multiply examples, but they will be familiar to most readers. Here I stress the most general point about *meanings*. STM is now a matter of debate – not because people fail to understand the issues, but because of the divergence of ethical-political frameworks within which the issues are 'read' and lived out. To understand contemporary reactions, the first desideratum is an understanding of that *variety* of perspectives. In this respect, *public* understanding of science, with the suggestion that the public (and science) is singular, is very misleading.

At the risk of seeming self-interested, I hope I have shown that one key discipline for broadening this perspective is *history*. As I argued above, the 'public' should be seen as comprising many 'groups' – some organised, some self-conscious, some merely showing common responses. These groups – be they profes-

sional, political or religious – change over time. To try to understand them without their histories may be as shallow as attempt-
ing a non-historical account of the National Health Service or of
Britain-in-Europe. But, alas, we have few such histories, just lots
of 'snapshots' that may not 'add up'.

To illuminate issues of public understanding we need more histories that are wide enough to include politics and the media, but
which also include the technical within the critical gaze. In as
much as historians have learned how to relate the technicalities
of investigation to wider questions of motivation, legitimation
and support, so they can offer a means by which their readers
can see STM as part of individual and collective *projects*. Much
the same end can sometimes be achieved through scientific biography or autobiography. The critical step, for historians, journalists or participants 'retelling', is the ability to *link* the
technical and the social, as projects which cross these realms.
That is one of the reasons why this book presents STM as a
variety of projects, as kinds of work.[8]

Such 'projects' may, of course, be much larger than any one
person's intentions. This is especially true in technoscience where
the overall aims may be negotiated so they correspond to no one
person's wishes, and where the project may give home to many
workers whose individual or collective aims may be at a tangent
to those of the project as a whole. Critical and constructive discussion of STM requires public access to reliable accounts of
how such scientific projects work – their dynamics and the
systems of control.

One of the distressing features of recent British debates has
been a tendency of 'modernisers' to conflate the advance of
knowledge, the progress of humankind and the interests of particular agricultural companies in one unholy trinity. As publics
we deserve better. We need more reliable histories of how such
multinationals work, how they handle adverse evidence and
respond to public pressures. We need analytical accounts of
recent history which distinguish between quality of life and gross
national products, and accounts of the increase of knowledge
which differentiate between kinds of knowledge and also

between the different ends for which knowledges are created, circulated and used. To all such accounts of projects, *meanings* are central. Even the most technical and 'pure' projects are built on *some* ethical and political principles, at minimum a belief in the private or public funding of scientific enquiry for its own sake.

Science, values and history

To recommend that STM be studied in terms of projects may also help make links with the excellent historical literature on science and religion, or more generally on science and value systems. Though formal religions and theology may now be marginal for the majority of literate populations in the West, they remain crucial in determining some aspects of world development (e.g. population policy), and they remain central to political life in much of the rest of the world. But even if we put aside the continuing roles of formal religions, we cannot ignore the questions once raised by science–religion debates. Those debates have shaped later attitudes and positions; so, for example, if you wish to understand the social and political orientations of British scientists and intellectuals in the twentieth century, you would do well to check out the religious denominations in which so many of them were raised (and which they may have reacted against). Many of our secular debates have their analogues in previous religious debates, some of which have been submitted to exemplary historical study. By thus comparing the present with the past, we may gain tools for the analysis of current issues, and a better appreciation of the 'metaphysical' commitments which underlie positions that now appear as non-religious or antireligious.

One could, in the manner of bioethics courses, simply debate the pros and cons of present positions, but history offers more (and not less). As well as using an analytical framework to clarify debates, it serves to link attitudes to traditions and social projects. It allows one to ask about the origins and pertinence of *questions* (as well as about answers). Why, for example, was vivisection a much bigger issue in later nineteenth-century Britain than in continental Europe? Why has it returned as a major issue (and what now is its geography)? What are the social dynamics of 'ecological activism' or of enthusiasm for nuclear energy? In such ways, historical study of political and ethical issues in STM may reorient students or wider publics to consider their own positions and responses historically – to see themselves as a part of history, if that is not too grand a claim. By trying to understand American fundamentalism, or the polemical activities of present-day critics of religion, they may gain an appreciation of the rootedness of attitudes in biography and history, and thus a more realistic appreciation of the roles and limits of public debates over vexed issues.

It is often useful to examine the interplay of a wide range of positions, e.g. on human evolution. The benefit of scholarship here is to show those various positions as occupied both by 'theologians' *and* 'scientists', and by many others who claimed no particular expertise. Scholars no longer write about the conflicts of 'science' with 'religion', but about conflicts between positions and parties *each of which*, in varying degrees, is *both* scientific and religious (or at least metaphysical – in the sense of accepting certain propositions as a basis for others). To take an example – later nineteenth-century debates about morals and evolution revolved around the supposed relationships between 'human values' and the 'facts' of human biology and history.

Herbert Spencer, the 'sociologist' friend of the novelist George Eliot, and the author of the phrase 'survival of the fittest', then saw evolution and human history as processes of differentiation – from simple animals to complex, from simple societies to the industrial West of his day. Imperial conquest and the subordination of primitive societies was both fact and value(d) – it was 'a good thing'; Spencer was a secularist, but much the same position was held by Christian evolutionists. By contrast, Alfred Russel Wallace, the plebeian naturalist who 'invented' evolution by natural selection independently of Darwin, refused to see human history as a continuation of natural selection; Victorian Britain was not evolution's goal. Wallace was a socialist and a spiritualist who insisted that the realm of human values be kept separate; that was part of the significance of his spiritualism. Somewhere between them, T. H. Huxley, Darwin's friend, came to moderate his early 'scientific naturalism', and to recognise more of a divide between facts and values, more of a separation between biology and ethics.[9]

All three positions have approximate counterparts in present debates about sociobiology. Some popularisers of human genetics talk as if we are prisoners of genetic factors established aeons ago. By contrast, most anthropologists treat questions of culture as independent of biology. Some philosophers have usefully picked apart the ways in which we might learn about human propensities – inherited and/or conditioned – without assuming that they are justifications (see Midgley, 1995). By analysing such debates over time, we can see how people have struggled with these shifting issues, in the form and with the resources that were 'provided' by their own contexts; we can see such struggles and debates as *rooted*. Here, of course, I am returning to the matters surveyed in the chapter on world-meanings; as a second and concluding example, closer to our time, we might consider the recent history of abortion.

It is now strange to recall that when abortion was discussed by American doctors in the 1950s, it was chiefly a matter of professional etiquette. Because more and more births were taking place in hospitals and pregnancies were supervised by hospital doctors, matters such as abortion were no longer quite so 'private' between the patient and her own doctor. Because other doctors and nurses were involved, and the institutions, so rules were called for. Professional committees debated the issues and the circumstances in which abortion was to be permitted. By the 1970s, the laws and practices had changed, but so, more fundamentally, had the conditions and the nature of the debate. The women's movement and the 1960s counter-culture more generally, had made abortion a question of women's rights. In turn, the Catholic Church had mobilised for the 'protection of the unborn'; Catholic women, at home and unused to politics, were organising, using telephones rather than public meetings (Luker,

1984). The issue remains fiercely contested in the USA, a matter on which presidential candidates must state their position and take the consequences. In Britain, abortion was legalised in 1967, and rapidly became, in effect, a matter of the woman's choice; it is still controversial, but accepted by the majority of the population.

Such histories, especially if they are *comparative*, may serve to help us understand the presentation of issues in our present, and the prospects of reaching accommodation, if not agreement. For whenever we approach issues which cut to the heart of our human concerns – be they abortion or AIDS, the preservation of environments or the prevention of hunger – the majority of intelligent commentators do in fact recognise the *variety of positions* held, each one involving both ethical principles and particular approaches to research and results. History can help us articulate and share these complexities, to contextualise them and perhaps to see something of their dynamics. History can bring STM (in)to life.

For in such matters, as so often in issues around 'science', we may need reminding to develop our everyday political understandings and to stop being hoodwinked by presentations of 'science' (or 'religion' or 'modernisation') as if they were *unitary* activities, with no choices as to values. To historicise science, technology and medicine is to find the human meanings of technical projects in their origins and in their development. That is the historical key to understanding present science, whether we are talking of esoteric research or of very public debates.

Science-technology-medicine, we may conclude, is far too varied to simply support or oppose. We have seen that *natural history* can be an instrument of domination and appropriation, or a celebration of 'nature's riches'. *Analysis* can 'deepen' or 'reduce' our understanding of the world; it can regulate or intensify our technical processes. *Experiment* is intrinsic to modern creativity, though it may sometimes be dangerous. The *technosciences* which bring to our homes the security of antibiotics and the riches of diverse human cultures, may also envelop us in a 'naturalisation' of commerce as powerful and seductive as the enchanted natures of the Renaissance (or of the medieval church which believed in saints as well as sponsors).

Since the seventeenth century, and especially since about 1800, STM has been a major component of historical change. There have been, and there will be, many 'scientific revolutions', as men and women continually interrogate and remake the world, and themselves. But as we have seen, women and men also remake themselves in 'cultural reformations' – by the interrogation and assertion of human values (which in turn help determine approaches in science).

Science, technology and medicine are now central to our economies and to our cultures; the politics of STM will be central to our futures. The public and private investments in commerce-led technoscience are now enormous. For that reason, we also need to invest critical intelligence and financial and political resources to ensure that public interests will be served as well as commercial interests. Understanding the history of STM is one way by which its products and processes can be opened to scrutiny and control by citizens and consumers.

Notes

1 I owe this observation to Simon Schaffer.
2 On drug regulation, see Temin (1980) and Abraham (1995); on embryos, see Mulkay (1997).
3 For an intriguing, hopeful account of contraceptive research for the Third World, see Oudshoorn (1998).
4 See essays by Tuchman, Lenoir and Holmes in Coleman and Holmes (1988).
5 On simulation in high-energy physics, see Galison (1997); on 'pro-grammes' as a new way of doing mathematics, see Hodgkin (1976).
6 For science education, see Brock (1990).
7 On the medical science that used physiology to focus on the statis-tically 'normal' rather than the individually 'natural', see the key study by Warner (1986); on the management of (chronic-pain) patients at the limits of modern medicine, see Baszanger (1998); and see the work of Canguilhem (1989) and Delaporte (1994).

8 For accessible 'case studies' of modern science and technology, stressing contingency and open-endedness, see Collins and Pinch (1993; 1998).

9 See Young (1985) for seminal discussions of the politics of evolution; and Bowler (1989a).

Bibliography

* Accessible works for those who are new to the history of science, technology and medicine.
** Useful reference work.

Aarsleff, H. (1983), *The Study of Language in England 1780–1860*, Minneapolis and London.

Abir-Am, P. G. (1997), 'The Molecular Transformation of Twentieth-Century Biology', in J. Krige and D. Pestre (eds), *Science in the Twentieth Century*, Amsterdam, 495–524.

Abraham, J. (1995), *Science, Politics and the Pharmaceutical Industry: Controversy and Bias in Drug Regulation*, London.

Ackerknecht, E. H. (1953), *Rudolf Virchow: Doctor, Statesman and Anthropologist*, Madison.

Ackerknecht, E. H. (1967), *Medicine at the Paris Hospital 1794–1848*, Baltimore.

Agar, J. (1998), *Science and Public Spectacle: The Work of Jodrell Bank in Post-War British Culture*, Amsterdam.

Albury, W. R. (1972), 'The Logic of Condillac and the Structure of French Chemical and Biological Theory, 1780–1801', unpublished PhD thesis, Johns Hopkins University, Baltimore.

Albury, W. R. (1977), 'Experiment and Explanation in the Physiology of Bichat and Magendie', *Studies in the History of Biology*, 1, 47–131.

Alder, K. (1997), *Engineering the Revolution: Arms and Enlightenment in France, 1763–1815*, Princeton.

* Allen, D. E. (1976), *The Naturalist in Britain: A Social History*, London.

Allen, G. E. (1978), *Life Science in the Twentieth Century*, Cambridge History of Science series, vol. 4, Cambridge and New York.

* Alpers, S. (1983), *The Art of Describing: Dutch Art in the Seventeenth Century*, Chicago.

Alter, P. (1987), *The Reluctant Patron: Science and the State in Britain 1850–1920*, Oxford.

Altick, R. D. (1978), *The Shows of London*, Cambridge, MA, and London.

Amsterdamska, O. and Hiddinga, A. (2000), 'The Analyzed Body', in R. Cooter and J. Pickstone (eds), *Medicine in the Twentieth Century*, Amsterdam.

Appel, T. A. (1987), *The Cuvier-Geoffroy Debate: French Biology in the Decades Before Darwin*, Oxford and New York.

Arnold, K. R. (1992), 'Cabinets for the Curious: Practicing Science in Early Modern English Museums', unpublished PhD thesis, Princeton University.

Arnold, M. (1965), *Culture and Anarchy (The Complete Prose Works of Matthew Arnold, Volume 5)*, Ann Arbor, MI (first published in 1869, London).

Ashplant, T. G. and Wilson, A. (1988), 'Present-Centred History and the Problem of Historical Knowledge', *Historical Journal*, 31, 253–74.

Ashworth, W. B. Jr (1990), 'Natural History and the Emblematic World View', in D. C. Lindberg and R. S. Westman (eds), *Reappraisals of the Scientific Revolution*, Cambridge, 303–32.

Atran, S. (1990), *Cognitive Foundations of Natural History: Towards an Anthropology of Science*, Cambridge.

Austoker, J. and Bryder, L. (eds) (1989), *Historical Perspectives on the Role of the MRC: Essays in the History of the Medical Research Council of the United Kingdom and its Predecessor, the Medical Research Committee, 1913–1953*, Oxford.

Babbage, C. (1832), *On the Economy of Machinery and Manufactures*, London.

Bachelard, G. (1938), *The Psychoanalysis of Fire*, trans. A. C. M. Ross, Boston.

* Barbour, I. G. (1966), *Issues in Science and Religion*, New York and London.

* Barnes, B. (1974), *Scientific Knowledge and Sociological Theory*, London.

Barnes, B. (1977), *Interests and the Growth of Knowledge*, London.

Barnes, B. (1982), *T. S. Kuhn and Social Science*, London.

* Basalla, G. (1988), *The Evolution of Technology*, Cambridge.

Baszanger, I. (1998), *Inventing Pain Medicine: From the Laboratory to the Clinic*, New Brunswick, NJ, and London.

Beer, J. J. (1959), *The Emergence of the German Dye Industry*, Urbana, IL.

* Ben-David, J. (1971), *The Scientist's Role in Society: A Comparative Study*, Englewood Cliffs, NJ.

Bendix, R. (1960), *Max Weber: An Intellectual Portrait*, New York.

Bennett, J. A. (1986), 'The Mechanics' Philosophy and the Mechanical Philosophers', *History of Science*, 24, 1–28.

Berg, M. (1980), *The Machinery Question and the Making of Political Economy 1815–1848*, Cambridge.

Berman, M. (1978), *Social Change and Scientific Organization: The Royal Institution, 1799–1844*, London.

Bernard, C. (1957), *An Introduction to the Study of Experimental Medicine*, trans. H. C. Greene, New York.

Berridge, V. (1996), *AIDS in the U.K.: The Making of Policy, 1981–1994*, Oxford.

* Berridge, V. (1999), *Health and Society in Britain Since 1939*, Cambridge.

Berridge, V. (2000), 'AIDS and Patient/Support Groups', in R. Cooter and J. Pickstone, (eds), *Medicine in the Twentieth Century*, Amsterdam.

Berthelot, M. (1879), *Synthèse Chimique*, 3rd edn, Paris.

Blume, S. (1992), *Insight and Industry: The Dynamics of Technical Change in Medicine*, Cambridge, MA, and London.

Blume, S. (2000), 'Medicine, Technology and Industry', in R. Cooter and J. Pickstone, (eds), *Medicine in the Twentieth Century*, Amsterdam.

Bowler, P. J. (1983), *The Eclipse of Darwinism: Anti-Darwinian Evolution Theories in the Decades around 1900*, Baltimore.

Bowler, P. J. (1988), *The Non-Darwinian Revolution: Reinterpreting a Historical Myth*, Baltimore.

* Bowler, P. J. (1989a), *Evolution: The History of an Idea*, rev. edn, Berkeley, CA.

Bowler, P. J. (1989b), *The Invention of Progress: The Victorians and the Past*, Oxford.

* Bowler, P. J. (1992), *The Fontana History of the Environmental Sciences*, London.

Bowler, P. J. (1996), *Life's Splendid Drama: Evolutionary Biology and the Reconstruction of Life's Ancestry, 1860–1940*, Chicago.

Bramwell, A. (1989), *Ecology in the 20th Century: A History*, New Haven, CT.

Braverman, H. (1974), *Labour and Monopoly Capitalism: The Degradation of Work in the Twentieth Century*, New York.

Bray, F. (1997), *Technology and Gender. Fabrics of Power in Late Imperial China*, Berkeley and Los Angeles.

Brock, W. H. (1990), 'Science Education', in R. C. Olby, G. N. Cantor, J. R. R. Christie and M. J. S. Hodge (eds), *Companion to the History of Modern Science*, London, 946–9.

* Brock, W. H. (1992), *The Fontana History of Chemistry*, London.

Brock, W. H. (1997), *Justus von Liebig: The Chemical Gatekeeper*, Cambridge.

Broman, T. H. (1996), *The Transformation of German Academic Medicine 1750–1820*, Cambridge.

Brooke, J. H. (1990), 'Science and Religion', in R. C. Olby, G. N. Cantor, J. R. R. Christie and M. J. S. Hodge (eds), *Companion to the History of Modern Science*, London, 763–82.

* Brooke, J. H. (1991), *Science and Religion: Some Historical Perspectives*, Oxford.

Browne, J. (1983), *The Secular Ark: Studies in the History of Biogeography*, New Haven, CT.

Browne, J. (1995), *Charles Darwin: Voyaging*, London.

Buchanan, R. A. (1989), *The Engineers: A History of the Engineering Profession in Britain 1750–1914*, London.

* Buchanan, R. A. (1992), *The Power of the Machine: The Impact of Technology from 1700 to the Present*, London.

Buchwald, J. Z. (1987), *The Rise of the Wave Theory of Light: Optical Theory and Experiment in the Early Nineteenth Century*, Chicago.

Bud, R. (1993), *The Uses of Life: A History of Biotechnology*, Cambridge.

Bud, R. (1998), 'Penicillin and the New Elizabethans', *British Journal for the History of Science*, 31, 305–33.

Bud, R. and Cozzens, S. E. (eds) (1992), *Invisible Connections: Instruments, Institutions, and Science*, Bellingham.

Bud, R. and Roberts, G. K. (1984), *Science versus Practice: Chemistry in Victorian Britain*, Manchester.

Burkhardt, R. W. Jr (1977), *The Spirit of System: Lamarck and Evolutionary Biology*, Cambridge, MA.

Butler, S. V. F. (1986), 'A Transformation of Training: The Formation of University Medical Faculties in Manchester, Leeds, and Liverpool, 1870–84', *Medical History*, 30, 115–31.

Butler, S. V. F. (1992), *Science and Technology Museums*, Leicester.

Butterfield, H. (1957), *The Origins of Modern Science 1300–1800*, London.

Bynum, W. F. (1993), 'Nosology', in W. F. Bynum and R. Porter (eds), *Companion Encyclopedia of the History of Medicine*, London and New York, 335–56.

* Bynum, W. F. (1994), *Science and the Practice of Medicine in the Nineteenth Century*, Cambridge.

** Bynum, W. F. and Porter, R. (eds), (1993) *Companion Encyclopedia of the History of Medicine*, London and New York.

Cahan, D. (1989), *An Institute of Empire: The Physikalische-Technische Reichsanstalt, 1871–1918*, Cambridge.

Canguilhem, G. (1975), *Etudes d'Histoire et de Philosophie des Sciences*, 3rd edn, Paris.

Canguilhem, G. (1989), *The Normal and the Pathological*, Cambridge, MA (French original, 1966).

Cannon, S. F. (1978), *Science in Culture: The Early Victorian Period*, New York.

Cantor, D. (1992), 'Cortisone and the Politics of Drama, 1949–55', in J. V. Pickstone (ed.), *Medical Innovations in Historical Perspective*, Basingstoke and London, 165–84.

Cantor, G. N. (1991), *Michael Faraday: Sandemanian and Scientist: A Study of Science and Religion in the Nineteenth Century*, Basingstoke.

Cantor, G., Gooding, D. and James, F. A. J. L. (1991), *Faraday*, Basingstoke.

Capshew, J. and Radar, K. (1992), 'Big Science: Price to the Present', *Osiris*, 7, 3–25.

* Cardwell, D. S. L. (1957), *The Organisation of Science in England*, London.

Cardwell, D. S. L. (1968), *John Dalton and the Progress of Science*, Manchester and New York.

Cardwell, D. S. L. (1971), *From Watt to Clausius: The Rise of Thermodynamics in the Early Industrial Age*, London.

Cardwell, D. S. L. (1989), *James Joule: A Biography*, Manchester and New York.

* Cardwell, D. S. L. (1994), *The Fontana History of Technology*, London.

Carlson, W. B. (1997), 'Innovation and the Modern Corporation: From Heroic Invention to Industrial Science', in J. Krige and D. Pestre (eds), *Science in the Twentieth Century*, Amsterdam, 203–26.

Carson, R. (1962), *Silent Spring*, Boston.

Cassirer, E. (1955), *The Philosophy of the Enlightenment*, Boston.

Channell, D. F. (1988), 'The Harmony of Theory and Practice: the Engineering Science of W.J.M. Rankine', *Technology and Culture*, 29, 98–103.

Chapman, W. R. (1985), 'Arranging Ethnology: A.H.L.F. Pitt Rivers and the Typological Tradition', in G. Stocking Jr (ed.), *Objects and Others*, Madison, 15–48.

Chevreul, M. E. (1824), *Considérations Générales sur l'Analyse Organique et sur ses Applications*, Paris.

Churchill, F. B. (1973), 'Chabry, Roux and the Experimental Method in Nineteenth-Century Embryology', in R. N. Giere and R. S. Westfall (eds), *Foundations of Scientific Method: The Nineteenth Century*, Bloomington.

Churchill, F. B. (1994), 'The Rise of Classical Descriptive Embryology', in S. F. Gilbert, (ed.), *A Conceptual History of Modern Embryology*, Baltimore and London, 1–29.

Clarke, A. E. (1998), *Disciplining Reproduction: Modernity, American Life Sciences, and 'the Problems of Sex'*, Berkeley, Los Angeles and London.

Clarke, E. and Jacyna, L. S. (1987), *Nineteenth-Century Origins of Neuroscientific Concepts*, Berkeley, CA.

Cohen, Y. (1997), 'Scientific Management and the Production Process', in J. Krige and D. Pestre (eds), *Science in the Twentieth Century*, Amsterdam, 111–25.

Coleman, W. (1964), *Georges Cuvier, Zoologist: A Study in the History of Evolutionary Theory*, Cambridge, MA.

Coleman, W. (1977), *Biology in the Nineteenth Century: Problems of Form, Function, and Transformation*, Cambridge and New York.

Coleman, W. and Holmes, F. L. (eds) (1988), *The Investigative Enterprise: Experimental Physiology in Nineteenth-Century Medicine*, Berkeley and Los Angeles.

Collingwood, R. G. (1938), *The Principles of Art*, London and Oxford.

* Collingwood, R. G. (1939), *An Autobiography*, Oxford.

Collingwood, R. G. (1945), *The Idea of Nature*, Oxford.

Collingwood, R. G. (1946), *The Idea of History*, Oxford.

Collins, H. (1992), *Changing Order: Replication and Induction in Scientific Practice*, Chicago and London.

* Collins, H. and Pinch, T. (1993), *The Golem: What Everyone Should Know about Science*, Cambridge.

* Collins, H. and Pinch, T. (1998), *The Golem at Large: What You Should Know about Technology*, Cambridge.

** Conrad, L. I., Neve, M., Porter, R. and Nutton, V. (eds) (1995), *The Western Medical Tradition, 800 BC to AD 1800*, Cambridge.

Cook, H. J. (1990),'The New Philosophy and Medicine in Seventeenth-Century England', in D. C. Lindberg and R. S. Westman (eds), *Reappraisals of the Scientific Revolution*, Cambridge, 397–436.

Cook, H. J. (1994), *Trials of an Ordinary Doctor: Joannes Groenevelt in Seventeenth-Century London*, Baltimore.

Cook, H. J. (1996), 'Physicians and Natural History', in N. Jardine, J. A. Secord and E. C. Spary (eds), *Cultures of Natural History*, Cambridge, 91–105.

Cooter, R. J. (1984), *The Cultural Meaning of Popular Science: Phrenology and the Organisation of Consent in Nineteenth-Century Britain*, Cambridge.

Cooter, R. J. (1993a), *Surgery and Society in Peace and War*, Basingstoke.

Cooter, R. J. (1993b), 'War and Modern Medicine', in W. F. Bynum and R. Porter (eds), *Companion Encyclopedia of the History of Medicine*, London and New York, 1536–73.

Cooter, R. J. (2000a), 'Disabled Body', in R. J. Cooter and J. V. Pickstone (eds), *Medicine in the Twentieth Century*, Amsterdam.

Cooter, R. J. (2000b), 'The Ethical Body', in R. J. Cooter and J. V. Pickstone (eds), *Medicine in the Twentieth Century*, Amsterdam.

** Cooter, R. J. and Pickstone, J. V. (eds) (2000), *Medicine in the Twentieth Century*, Amsterdam.

Cornwell, J. (1984), *Hard Earned Lives. Accounts of Health and Illness in East London*, London.

Corsi, P. (1988), *The Age of Lamarck: Evolutionary Theories in France, 1790–1830*, Berkeley, CA.

Cowan, R. S. (1983), *More Work for Mother: The Ironies of Household Technology from the Open Hearth to the Microwave*, New York.

* Cowan, R. S. (1997), *A Social History of American Technology*, New York and Oxford.

Crosland, M. P. (1967), *The Society of Arceuil: A View of French Science at the Time of Napoleon I*, London.

Crosland, M. P. (1970), 'Berthelot', *Dictionary of Scientific Biography*, vol. 2, New York, 63–72.

Crosland, M. P. (1978), *Historical Studies in the Language of Chemistry*, 2nd edn, New York.

Crosland, M. P. (1980), 'Chemistry and the Chemical Revolution', in G. S. Rousseau and R. S. Porter (eds), *The Ferment of Knowledge*, Cambridge, 389–416.

Crosland, M. P. and Smith, C. (1978), 'The Transmission of Physics from France to Britain: 1800–1840', *Historical Studies in Physical Sciences*, 9, 1–61.

Crowther, J. G. (1974), *The Cavendish Laboratory, 1874–1974*, New York.

Cullen, W. (1816), *First Lines of the Practice of Physic*, Philadelphia.

Cunningham, A. (1997), *The Anatomical Renaissance: The Resurrection of the Anatomical Projects of the Ancients*, Aldershot and Brookfield, VT.

Cunningham, A. and Jardine, N. (eds) (1990), *Romanticism and the Sciences*, Cambridge and New York.

Cunningham, A. and Williams, P. (1993), 'De-centring the "Big Picture": The *Origins of Modern Science* and the Modern Origins of Science', *British Journal for the History of Science*, 26, 407–32.

Currer, C. and Stacey, M. (eds) (1986), *Concepts of Health, Illness and Disease*, Leamington Spa, Hamburg and New York.

Daniels, G. H. (1967), 'The Process of Professionalisation in American Science: The Emergent Period, 1820–1860', *Isis*, 58, 151–68.

Darwin, C. (1859), *The Origin of Species*, 1982 edn, Harmondsworth.

Daston, L. and Park, K. (1998), *Wonders and the Order of Nature 1150–1750*, New York.

Daudin, H. (1926), *De Linné à Jussieu. Méthodes de la Classification et Idée de Série en Botanique et en Zoologie (1740–1790)*, Paris.

Daudin, H. (1926), *Cuvier et Lamarck, Les Classes Zoologiques et l'Idée de Série Animale (1790–1830)*, Paris.

Davenport-Hines, R. and Slinn, J. A. (1992), *Glaxo: A History to 1962*, Cambridge.

Davies, P. (1983), 'Sir Arthur Schuster, 1851–1934', unpublished PhD thesis, UMIST.

Debus, A. G. (1978), *Man and Nature in the Renaissance*, Cambridge History of Science series, vol. 6, Cambridge and New York.

Delaporte, F. (ed.) (1994), *A Vital Rationalist: Selected Writings from Georges Canguilhem*, trans. A. Goldhammer, New York.

Dennis, M. A. (1987), 'Accounting for Research; New Histories of Corporate Laboratories and the Social History of American Science', *Social Studies in Science*, 17, 479–518.

Desmond, A. (1976), *The Hot-Blooded Dinosaurs: A Revolution in Palaeontology*, New York.

Desmond, A. (1989), *The Politics of Evolution: Morphology, Medicine and Reform in Radical London*, Chicago and London.

Desmond, A. (1994), *Huxley: The Devil's Disciple*, London.

Desmond, A. (1997), *Huxley: Evolution's High Priest*, London.

* Desmond, A. and Moore, J. R. (1992), *Darwin*, London.

Dhombres, N. and J. (1989), *Naissance d'un Nouveau Pouvoir: Sciences et Savants en France (1793–1824)*, Paris.

Dibner, B. (1972), 'Hopkinson', *Dictionary of Scientific Biography*, vol. 6, New York, 504.

Dolman, C. E. (1971), 'Ehrlich', *Dictionary of Scientific Biography*, vol. 4, New York, 295–305.

Donovan. A. L. (1975), *Philosophical Chemistry in the Scottish Enlightenment: The Doctrines and Discoveries of William Cullen and Joseph Black*, Edinburgh.

Donovan, A. L. (1996), *Antoine Lavoisier: Science, Administration and Revolution*, Cambridge.

Dorson, R. M. (1968), *The British Folklorists: A History*, London.

Dostrovsky, S. (1976), 'Wheatstone', *Dictionary of Scientific Biography*, vol. 14, New York, 289–91.

Duden, B. (1991), *The Woman Beneath the Skin: A Doctor's Patients in Eighteenth-Century Germany*, Cambridge, MA.

Duffin, J. (1998), *To See with a Better Eye: A Life of RTH Laennec*, Princeton, NJ.

Dunsheath, P. (1962), *A History of Electrical Engineering*, London.

Dupree, A. H. (1957), *Science in the Federal Government: A History of Policies and Activities to 1940*, Cambridge, MA.

Eagleton, T. (1983), *Literary Theory: An Introduction*, Oxford.

Edgerton, D. E. H. (1992), *England and the Aeroplane: An Essay on a Militant and Technological Nation*, Basingstoke.

Edgerton, D. E. H. (1996a), *Industrial Research and Innovation in Business*, Brookfield, VT.

* Edgerton, D. E. H. (1996b), *Science, Technology and the British Industrial 'Decline' 1870–1970*, Cambridge.

Edgerton, D. E. H. (1997), 'Science in the United Kingdom: A Study in the Nationalization of Science', in J. Krige and D. Pestre (eds), *Science in the Twentieth Century*, Amsterdam, 759–76.

Edgerton, D. E. H. (1999), 'From Innovation to Use: Ten Eclectic Theses on the Historiography of Technology', *History and Technology*, 16, 111–36.

Edgerton, D. E. H. and Horrocks, S. (1994), 'British Industrial

R&D before 1945', *Economic History Review*, 47, 213–38.

Elsner, J. and Cardinal, R. (eds) (1994), *The Cultures of Collecting*, London.

Entralgo, P. L. (1969), *Doctor and Patient*, London.

Eribon, D. (1991), *Michel Foucault*, trans. B. Wing, Cambridge, MA.

* Evans, R. J. (1987), *Death in Hamburg*, Oxford.

Farley, J. (1991), *Bilharzia: A History of Imperial Tropical Medicine*, Cambridge.

Farrar, W. V. (1997), *Chemistry and the Chemical Industry in the Nineteenth Century: The Henrys of Manchester and other Studies*, ed. by R. L. Hills and W. H. Brock, Aldershot.

Faulkner, W. and Kerr, E. A. (1997), 'On Seeing Brockenspectres: Sex and Gender in Twentieth-Century Science', in J. Krige and D. Pestre (eds), *Science in the Twentieth Century*, Amsterdam, 43–60.

Ferraris, M. (1996), *History of Hermeneutics*, trans. L. Somigli, Atlantic Highlands, NJ.

Feyerabend, P. (1975), *Against Method: Outline of an Anarchistic Theory of Knowledge*, London.

Figlio, K. (1977), 'The Historiography of Scientific Medicine: An Invitation to the Human Sciences', *Comparative Studies in Science and Society*, 19, 262–86.

Findlen, P. (1994), *Possessing Nature: Museums, Collecting and Scientific Culture in Early Modern Italy*, Berkeley, CA.

Fisher, N. (1990), 'The Classification of the Sciences', in R. C. Olby, G. N. Cantor, J. R. R. Christie and M. J. S. Hodge (eds), *Companion to the History of Modern Science*, London, 853–68.

Fissell, M. E. (1991), *Patients, Power and the Poor in Eighteenth-Century Bristol*, Cambridge.

Forgan, S. (1994), 'The Architecture of Display: Museums, Universities and Objects in Nineteenth-Century Britain', *History of Science*, 32, 139–62.

Foster, W. D. (1961), *A Short History of Clinical Pathology*, London.

Foster, W. D. (1983), *Pathology as a Profession in Great Britain*, London.

Foucault, M. (1970), *The Order of Things. An Archaeology of the Human Sciences*, London.

Foucault, M. (1971), *Madness and Civilisation*, London.

Foucault, M. (1972), *The Archaeology of Knowledge*, trans. A. M. Sheridan, London.

Foucault, M. (1973), *The Birth of the Clinic*, London.

Foucault, M. (1979), *Discipline and Punish*, trans. A. Sheridan, London.

Fox, R. (1992), *The Culture of Science in France, 1700–1900*, Aldershot, and Brookfield, VT.

Fox, R. (ed.) (1996), *Technological Change: Methods and Themes in the History of Technology*, Amsterdam.

Fox, R. and Guagnini, A. (eds) (1993), *Education, Technology and Industrial Performance in Europe, 1850–1939*, Cambridge.

Fox, R. and Guagnini, A. (1998), 'Laboratories, Workshops and Sites. Concepts and Practices of Research in Industrial Europe, 1800–1914, part 1', *Historical Studies in the Physical and Biological Sciences*, 29:1, 55–139.

Fox, R. and Guagnini, A. (1999), 'Laboratories, Workshops and Sites. Concepts and Practices of Research in Industrial Europe, 1800–1914, part 2', *Historical Studies in the Physical and Biological Sciences*, 29:2, 191–294.

Fox, R. and Weisz, G. (1980), *The Organisation of Science and Technology in France 1808–1914*, Cambridge.

Frängsmyr, T. (ed.) (1984), *Linnaeus: The Man and his Work*, Berkeley, CA.

French, R. (1994), *William Harvey's Natural Philosophy*, Cambridge.

French, R. and Wear, A. (eds) (1991), *British Medicine in An Age of Reform*, London.

French, R. D. (1975), *Antivivisection and Medical Science in Victorian Society*, Princeton, NJ.

Galambos, L. and Sturchio, J. L. (1997), 'The Transformation of the Pharmaceutical Industry in the Twentieth Century', in J. Krige and D. Pestre (eds), *Science in the Twentieth Century*, Amsterdam, 227–52.

Galison, P. L. (1987), *How Experiments End*, Chicago and London.

Galison, P. L. (1997), *Image and Logic: A Material Culture of Microphysics*, Chicago and London.

Galison, P. L. and Assmus, A. (1989), 'Artificial Clouds, Real Particles', in D. Gooding, T. Pinch and S. Schaffer (eds) *The Uses of Experiment: Studies in the Natural Sciences*, Cambridge, 225–74.

Galison, P. L. and Hevly, B. (eds) (1992), *Big Science: The Growth of Large-Scale Research*, Stanford, CA.

Gaudillière, J.-P. and Löwy, I. (eds) (1998), *The Invisible Industrialist: Manufactures and the Production of Scientific Knowledge*, Basingstoke, London and New York.

Geiger, R. (1992), 'Science, Universities and National Defense, 1945–1970', *Osiris*, 7, 226–48.

Geiger, R. L. (1997), 'Science and the University: Patterns from the US Experience in the Twentieth Century', in J. Krige and D. Pestre (eds), *Science in the Twentieth Century*, Amsterdam, 159–74.

Geison, G. L. (1970–80), 'Pasteur', *Dictionary of Scientific Biography*, vol.10, New York, 350–416

Geison, G. L. (1978), *Michael Foster and the Cambridge School of Physiology*, Princeton, NJ.

Geison, G. L. (ed.) (1984), *Professions and the French State, 1700–1900*, Philadelphia.

Geison, G. L. (1995), *The Private Science of Louis Pasteur*, Princeton, NJ.

Gelfand, T. (1980), *Professionalising Modern Medicine: Paris Surgeons and Medical Science and Institutions in the Eighteenth Century*, Westport, CT.

Gibson, A. and Farrar, W. V. (1974), 'Robert Angus Smith, F.R.S. and "Sanitary Science"', *Notes and Records, Royal Society of London*, 28, 241–62.

Gilbert, S. F. (ed.) (1994), *A Conceptual History of Modern Embryology*, Baltimore.

Gilfillan, S. C. (1962), *The Sociology of Invention*, Chicago.

Gill, S. (1989), *William Wordsworth. A Life*, Oxford.

Gillispie, C. C. (1965), 'Science and Technology', in C. W. Crawley (ed.) *The New Cambridge Modern History*, vol. 9, Cambridge, 118–45.

Gillispie, C. C. (1967), *The Edge of Objectivity*, Princeton, NJ.

Gillispie, C. C. (1971), *Lazare Carnot, Savant*, Princeton, NJ.

Gillispie, C. C. (1972), 'The Natural History of Industry', in A. E. Musson (ed.), *Science, Technology and Economic Growth in the Eighteenth Century*, London, 121–35.

Gillispie, C. C. (1980), *Science and Polity in France at the End of the Old Regime*, Princeton, NJ.

Golinski, J. (1992), *Science as Public Culture: Chemistry and Enlightenment in Britain, 1760–1820*, Cambridge.

* Golinski, J. (1998), *Making Natural Knowledge: Constructivism and the History of Science*, Cambridge.

Gooday, G. (1990), 'Precision Measurement and the Genesis of Physics Teaching Laboratories in Victorian Britain', *British Journal for the History of Science*, 23, 25–51.

Gooday, G. (1991a), 'Teaching Telegraphy and Electrotechnics in the Physics Laboratory: William Ayrton and the Creation of an Academic Space for Electrical Engineering, 1873–84', *History of Technology*, 13, 73–111.

Gooday, G. (1991b), '"Nature", in the Laboratory: Domestication and Discipline with the Microscope in Victorian Life Science', *British Journal for the History of Science*, 24, 307–41.

Gooday, G. (1995), 'The Morals of Energy Metering: Conducting and Deconstructing the Precision of the Victorian Electrical Engineer's Ammeter and Voltmeter', in M. N. Wise (ed.), *The Values of Precision*, Princeton, NJ, 230–82.

Gooding, D. (1985), '"In Nature's School": Faraday as an Experimentalist', in D. Gooding and F. A. J. L. James (eds), *Faraday Rediscovered*, New York, 105–35.

Gooding, D. and James, F. A. J. L. (eds) (1985), *Faraday Rediscovered*, New York.

Gooding, D., Pinch, T. and Schaffer, S. (eds) (1989), *The Uses of Experiment: Studies in the Natural Sciences*, Cambridge.

Goodman, J. (2000), 'Pharmaceutical Industry', in R. Cooter and J. Pickstone (eds), *Medicine in the Twentieth Century*, Amsterdam.

Gough, J. B. (1970), 'Becquerel (1820–1891)', *Dictionary of Scientific Biography*, vol. 1, New York, 555–6.

Gouk, P. (1997), 'Natural Philosophy and Natural Magic', in E. Fucíková (ed.), *Rudolf II and Prague: The Court and the City*, London, 231–7.

Gould, P. C. (1988), *Early Green Politics: Back to Nature, Back to the Land, and Socialism, 1880–1900*, London.

* Gould, S. J. (1978), *Ontogeny and Phylogeny*, Cambridge, MA.

* Gould, S. J. (1987), *Time's Arrow, Time's Cycle*, Cambridge, MA.

Gowing, M. (1964), *Britain and Atomic Energy, 1939–1945*, Basingstoke.

Grafton, A. (1992), *New Worlds, Ancient Texts: The Power of Tradition and the Shock of Discovery*, Cambridge, MA.

Granshaw, L. (1985), *St Mark's Hospital, London: A Social History of a Specialist Hospital*, London.

Grattan-Guiness, I. (1981), 'Mathematical Physics in France, 1800–1835', in H. N. Jahnke and M. Otte (eds), *Epistemological and Social Problems of the Sciences in the Early Nineteenth Century*, Dordrecht, 349–70.

Grattan-Guiness, I. and Ravetz, J. R. (1972), *Joseph Fourier, 1769–1830: A Survey of his Life and Work based on a Critical Edition of the Monograph on the Propagation of Heat*, Cambridge, MA.

Greenaway, F. (1966), *John Dalton and the Atom*, London.

Grmek, M. D. (1970), 'Claude Bernard', *Dictionary of Scientific Biography*, vol. 2, New York, 24–34.

Guagnini, A. (1991), 'The Fashioning of Higher Technical Education and Training in Britain: The Case of Manchester, 1851–1914', in H. F. Gospel (ed.), *Industrial Training and Technological Innovation: A Comparative Historical Study*, London, 69–92.

Gutting, G. (1989), *Michel Foucault's Archaeology of Scientific Reason*, Cambridge.

Gutting, G. (1990), 'Continental Philosophy and the History of Science', in R. C. Olby, G. N. Cantor, J. R. R. Christie and M. J. S. Hodge (eds), *Companion to the History of Modern Science*, London, 127–47.

Haber, L. F. (1958), *The Chemical Industry during the Nineteenth Century: A Study of the Economic Aspect of Applied Chemistry in Europe and North America*, Oxford.

Habermas, J. (1989), *The Structural Transformation of the Public Sphere: An Inquiry into a Category of Bourgeois Society (1962)*, Cambridge, MA.

Hacking, I. (1983), *Representing and Intervening: Introductory Topics in the Philosophy of Natural Science*, Cambridge.

Hacking, I. (1990), *The Taming of Chance*, Cambridge.

Haigh, E. (1984), 'Xavier Bichat and the Medical Theory of the Eighteenth Century', *Medical History*, supplement no. 4, London.

Haines, B. (1978), 'The Inter-Relations between Social, Biological and Medical Thought, 1750–1850: Saint-Simon and Comte', *British Journal for the History of Science*, 11, 19–35.

Hallam, A. (1973), *A Revolution in the Earth Sciences: From Continental Drift to Plate Tectonics*, Oxford.

Hamlin, C. (1998), *Public Health and Social Justice in the Age of Chadwick: Britain, 1800-1854*, Cambridge and New York.

Hankins, T. L. (1985), *Science and the Enlightenment*, Cambridge History of Science series, vol. 8, Cambridge and New York.

Haraway, D. (1989), *Primate Visions: Gender, Race and Nature in the World of Modern Science*, London and New York.

Hardie, D. F. W. and Pratt, J. D. (1966), *A History of the Modern British Chemical Industry*, Oxford.

Harding, S. (1991), *Whose Science? Whose Knowledge? Thinking From Women's Lives*, Milton Keynes.

Harding, S. and O'Barr, J. F. (eds) (1987), *Sex and Scientific Enquiry*, Chicago and London.

Harris, H. (1998), *The Birth of the Cell*, New Haven, CT, and London.

Harvey, A. M. (1981), *Science at the Bedside; Clinical Research in American Medicine, 1905–1945*, Baltimore.

Harvey, W. (1628), *On the Motion of the Heart and of the Blood*, many editions.

Harwood, J. (1993), *Styles of Scientific Thought: the German Genetics Community, 1900–1933*, Chicago.

Heilbron, J. L. (1979), *Electricity in the 17th and 18th Centuries: A Study of Early Modern Physics*, Berkeley, CA, and London.

Helman, C. G. (1986), '"Feed a Cold, Starve a Fever": Folk Models of Infection in an English Suburban Community, and their Relation to Medical Treatment', in C. Currer and M. Stacey (eds), *Concepts of Health, Illness and Disease*, Leamington Spa, Hamburg and New York, 211–31.

Henry, J. (1990), 'Magic and Science in the Sixteenth and Seventeenth Centuries', in R. C. Olby, G. N. Cantor, J. R. R. Christie and M. J. S. Hodge (eds), *Companion to the History of Modern Science*, London, 583–96.

* Henry, J. (1997), *The Scientific Revolution and the Origins of Modern Science*, Basingstoke.

Herrlinger, R. (1970), *History of Medical Illustration: From Antiquity to A.D. 1600*, trans G. F. Uitgeverig and N. V. Callenbach, London.

Herschel, J. (1835), *Preliminary Discourse on the Study of Natural Philosophy* (first published in 1830), 1999 edn, Bristol.

Hills, R. L. (1989), *Power from Steam: A History of the Stationary Steam Engine*, Cambridge.

Hobsbawm, E. J. (1962), *The Age of Revolution Europe 1789–1848*, London.

Hobsbawm, E. J. (1975), *The Age of Capital 1848–1875*, London.

Hodgkin, L. (1976), 'Politics and Physical Sciences', *Radical Science Journal*, 4, 29–60.

Holmes, F. L. (1974), *Claude Bernard and Animal Chemistry*, Cambridge, MA.

Holmes, F. L. (1985), *Lavoisier and the Chemistry of Life*, Madison.

Homburg, E. (1992), 'The Emergence of Research Laboratories in the Dyestuffs Industry, 1870–1900', *British Journal for the History of Science*, 25, 91–112.

Hooper-Greenhill, E. (1992), *Museums and the Shaping of Knowledge*, London.

Hopwood, N. (forthcoming), 'Embryology', in P. Bowler and J. V. Pickstone (eds), *Life and Earth Sciences*, Cambridge History of Sciences, vols 19–20.

Hoskin, M. A. (1970), 'Aitken', *Dictionary of Scientific Biography*, vol. 1, New York, 87–8.

Hounshell, D. A. and Smith, J. K. Jr (1988), *Science and Corporate Strategy: Dupont R & D, 1902–1980*, Cambridge.

Hufbauer, K. (1982), *The Formation of the German Chemical Community*, Berkeley, CA.

Hughes, H. S. (1974), *Consciousness and Society*, London (first published 1959, St Albans).

Hughes, J. (1998), 'Plasticine and Valves: Industry, Instrumentation and the Emergence of Nuclear Physics', in J.-P. Gaudillière and I. Löwy (eds), *The Invisible Industrialist: Manufactures and the Production of Scientific Knowledge*, Basingstoke, London and New York, 58–101.

Hughes, T. P. (1983), *Networks of Power: Electrification in Western Society 1880–1930*, Baltimore.

* Hughes, T. P. (1989), *American Genesis: A Century of Invention and Technical Enthusiasm 1870–1970*, New York.

Hughes, T. P. (1996), 'Managing Complexity: Interdisciplinary Advisory Committees', in R. Fox (ed.), *Technological Change*, Amsterdam, 229–245.

Inkster, I. (1997), *Scientific Culture and Urbanisation in Industrialising Britain*, Aldershot.

Inkster, I. and Morrell, J. (eds) (1981), *Metropolis and Province: Science in British Culture, 1780–1850*, London.

Irwin, A. and Wynne, B. (eds) (1996), *Misunderstanding Science? The Public Reconstruction of Science and Technology*, Cambridge.

Israel, P. (1992), *From Machine Shop to Industrial Laboratory: Telegraphy and the Changing Context of American Invention, 1830–1920*, Baltimore, and London.

Jacob, F. (1974), *The Logic of Living Systems: A History of Heredity*, trans. B. E. Spillmann, London.

* Jacob, F. (1988), *The Logic of Life*, London.

Jacob, M. C. (1997), *Scientific Culture and the Making of the Industrial West*, Oxford.

Jahnke, H. N. and Otte, M. (eds) (1981), *Epistemological and Social Problems of the Sciences in the Early Nineteenth Century*, Dordrecht.

Jankovic, V. (2000), *Reading the Skies: A Cultural History of the English Weather, 1660–1820*, Manchester and Chicago.

Jardine, L. (1999), *Ingenious Pursuit: Building the Scientific Revolution*, London.

** Jardine, N., Secord, J. A. and Spary, E. C. (eds) (1995), *Cultures of Natural History*, Cambridge and New York.

Jewson, N. D. (1974), 'Medical Knowledge and the Patronage System in Eighteenth-Century England', *Sociology*, 8, 369–85.

Jewson, N. D. (1976), 'The Disappearance of the Sick Man from Medical Cosmology', *Sociology*, 10, 225–44.

Johnson, J. A. (1990), *The Kaiser's Chemists: Science and Modernisation in Imperial Germany*, London.

Jordanova, L. (1984), *Lamarck*, Oxford.

Josephson, M. (1959), *Edison*, New York.

Jungnickel, C. and McCormmach, R. (1986), *Intellectual Mastery of Nature: Theoretical Physics from Ohm to Einstein*, vol. 1, *The Torch of Mathematics, 1800–1870*, Chicago.

Kargon, R. H. (1977), *Science in Victorian Manchester: Enterprise and Expertise*, Manchester.

Kauffman, G. B. (1974), 'Magnus, Heinrich Gustav, *Dictionary of Scientific Biography*, vol. 9, New York, 18–19.

Keller, E. F. (1983), *A Feeling for the Organism: The Life and Work of Barbara McClintock*, San Francisco.

Keller, E. F. and Longino, H. E. (1996), *Feminism and Science*, Oxford and New York.

Kemp, M. (1990), *The Science of Art: Optical Themes in Western Art from Brunelleschi to Seur*, New Haven, CT, and London.

Kemp, M. (1997), *Behind the Picture: Art and Evidence in the Italian Renaissance*, New Haven, CT, and London.

* Kevles, D. J. (1985), *In the Name of Eugenics: Eugenics and the Uses of Human Heredity*, New York.

Kevles, D. J. (1987), *The Physicists: The History of a Scientific Community in Modern America*, Cambridge, MA.

Kevles, D. J. (1998), *The Baltimore Case: A Trial of Politics, Science and Character*, New York and London.

Klein, N. (2000), *No Logo: Taking Aim at the Brand Bullies*, London.

Knight, D. M. (1967), *Atoms and Elements*, London.

Knight, D. M. (1970), 'Antoine-César Becquerel (1788–1878)', *Dictionary of Scientific Biography*, vol. 1, New York, 557–8.

Knight, D. M. (1986), *The Age of Science: The Scientific World View in the Nineteenth Century*, Oxford.

Knight, D. M. (1992), *Ideas in Chemistry: A History of the Science*, London.

Knoepflmacher, U. C. and Tennyson, G. B. (1977), *Nature and the Victorian Imagination*, Berkeley and Los Angeles.

* Kragh, H. (1987), *An Introduction to the Historiography of Science*, Cambridge.

Krausz, M. (ed.) (1972), *Critical Essays on the Philosophy of R.G. Collingwood*, Oxford.

Krementsov, N. (1997), 'Russian Science in the Twentieth Century', in J. Krige and D. Pestre (eds), *Science in the Twentieth Century*, Amsterdam, 777–94.

Kremer, R. L. (1992), 'Building Institutes for Physiology in Prussia, 1836–1846: Contexts, Interests and Rhetoric', in A. Cunningham and P. Williams (eds), *The Laboratory Revolution in Medicine*, Cambridge, 72–109.

Krige, J. (1997), 'The Politics of European Scientific Collaboration', in J. Krige and D. Pestre (eds), *Science in the Twentieth Century*, Amsterdam, 897–918.

** Krige, J. and Pestre, D. (eds) (1997), *Science in the Twentieth Century*, Amsterdam.

Kuhn, T. S. (1957), *The Copernican Revolution: Planetary Astronomy in the Development of Western Thought*, New York.

* Kuhn, T. S. (1962), *The Structure of Scientific Revolutions*, Chicago.

Kuhn, T. S. (1977), *The Essential Tension: Selected Studies in Scientific Tradition and Change*, Chicago and London.

Kuhn, T. S. (1977a), 'Energy Conservation as an Example of Simultaneous Discovery', *The Essential Tension: Selected Studies in Scientific Tradition and Change*, Chicago and London, 66–104.

Kuhn, T. S. (1977b), 'Mathematical versus Experimental Traditions in the Development of Physical Science', *The Essential Tension*, Chicago and London, 311–65.

Landes, D. S. (1969), *The Unbound Prometheus: Technological Change and Industrial Development in Western Europe from 1750 to the Present*, Cambridge.

Lanz, J. M. and de Betancourt, A. (1808), *Essai sur la Composition de Machines*, Paris.

Larson, J. L. (1971), *Reason and Experience: The Representation of Natural Order in the Work of Carl von Linné*, Berkeley, CA.

Latour, B. (1987a), *The Pasteurization of France*, Cambridge, MA.

* Latour, B. (1987b), *Science in Action*, Milton Keynes.

Latour B. (1993), *We Have Never Been Modern*, London.

Laudan, R. (1987), *From Mineralogy to Geology: The Foundations of Science, 1650–1830*, Chicago.

Laudan, R. (1990), 'The History of Geology, 1780–1840', in R. C.

Olby, G. N. Cantor, J. R. R. Christie and M. J. S. Hodge (eds), *Companion to the History of Modern Science*, London, 314–25.

Lawrence, C. (1985), 'Incommunicable Knowledge: Science, Technology and the Clinical Art in Britain, 1850–1914', *Journal of Contemporary History*, 10, 503–20.

* Lawrence, C. (1994), *Medicine in the Making of Modern Britain, 1700–1920*, London and New York.

Lawrence, C. (1997), 'Clinical Research', in J. Krige and D. Pestre (eds), *Science in the Twentieth Century*, Amsterdam, 439–59.

Lawrence, C. and Weisz, G. (eds) (1998), *Greater than the Parts: Biomedicine 1920–1950*, Oxford.

Lawrence, S. (1996), *Charitable Knowledge: Hospital Pupils and Practitioners in Eighteenth-Century London*, Cambridge and New York.

Layton, D., Jenkins, E., Macgill, S. and Davey, A. (1993), *Inarticulate Science? Perspectives on the Public Understanding of Science and some Implications for Science Education*, Driffield, East Yorkshire.

Layton, E. (1971), 'Mirror Image Twins: The Communities of Science and Technology in 19th Century America', *Technology and Culture*, 12, 562–80.

Layton, E. (1974), 'Technology as Knowledge', *Technology and Culture*, 15, 31–41.

Lecourt, D. (1975), *Marxism and Epistemology: Bachelard, Canguilhem and Foucault*, trans. B. Brewster, London.

Lenoir, T. (1982), *The Strategy of Life: Teleology and Mechanics in Nineteenth-Century German Biology*, Dordrecht and London.

Lenoir T. (1988), 'A Magic Bullet: Research for Profit and the Growth of Knowledge in Germany around 1900', *Minerva*, 26, 66–88.

Lenoir T. (1997), *Instituting Science: The Cultural Production of Scientific Disciplines*, Stanford, CA.

Lepenies, W. (1985), *Between Literature and Science: The Rise of Sociology*, trans., R. J. Hollingdale, Cambridge and Paris.

Lesch, J. E. (1984), *Science and Medicine in France: The Emergence of Experimental Physiology 1790–1855*, Cambridge, MA.

Lesch, J. E. (1988), 'The Paris Academy of Medicine and Experimental Science, 1820–1848', in W. Coleman and F. L. Holmes (eds), *The Investigative Enterprise*, Berkeley and Los Angeles, 100–138.

Levine, J. P. (1977), *Dr. Woodward's Shield: History, Science, and Satire in Augustan England*, Ithaca, NY, and London.

Levine, P. (1986), *The Amateur and the Professional: Antiquarians, Historians and Archaeologists in Victorian England, 1838–1886*, Cambridge.

Lewis, C. S. (1954), *English Literature in the Sixteenth Century, Excluding Drama*, Oxford.

Liebenau, J. (1987), *Medical Science and Medical Industry: The Formation of the American Pharmaceutical Industry*, Basingstoke.

Lindberg, D. C. and Westman, R. S. (eds) (1990), *Reappraisals of the Scientific Revolution*, Cambridge.

Lovell, B. (1976), *P.M.S. Blackett: A Biographical Memoir*, London.

Löwy, I. (1996), *Between Bench and Bedside: Science, Healing, and Interleukin-2 in a Cancer Ward*, Cambridge, MA, and London.

Löwy, I. (1997), 'Cancer: The Century of the Transformed Cell', in J. Krige and D. Pestre (eds), *Science in the Twentieth Century*, Amsterdam, 461–77.

Löwy, I. (2000), 'The Experimental Body', in R. Cooter and J. Pickstone (eds), *Medicine in the Twentieth Century*, Amsterdam.

Luker, K. (1984), *Abortion and the Politics of Motherhood*, Berkeley, CA, and London.

Lundgren, P. (1990), 'Engineering Education in Europe and the USA, 1750–1930: The Rise to Dominance of School Culture and the Engineering Professions', *Annals of Science*, 47, 33–75.

Macdonald, M. (1981), *Mystical Bedlam: Madness, Anxiety and Healing in Seventeenth-Century England*, Cambridge.

Macdonald, M. (1990), *Sleepless Souls: Suicide in Early Modern England*, Oxford.

Macfarlane, G. (1979), *Howard Florey: The Making of a Great Scientist*, Oxford.

Macfarlane, G. (1984), *Alexander Fleming: The Man and the Myth*, London.

Mackay, A. (1984), *The Making of the Atomic Age*, Oxford.

MacKenzie, J. M. (1988), *The Empire of Nature: Hunting, Conservation, and British Imperialism*, Manchester.

MacLeod, C. (1996), 'Concepts of Invention and the Patent Controversy in Victorian Britain', in R. Fox (ed.), *Technological Change*, Amsterdam, 137–53.

MacLeod, R. (1976), 'Tyndall', *Dictionary of Scientific Biography*, vol. 13, New York, 521–4.

MacLeod, R. (1996), *Public Science and Public Policy in Victorian England*, Aldershot.

Magnello, E. (2000), *The National Physical Laboratory: An Illustrated History*, London.

Mahoney, M. S. (1997), 'The Search for a Mathematical Theory', in J. Krige and D. Pestre (eds), *Science in the Twentieth Century*, Amsterdam, 617–34.

Maienschein, J. (1991), *Transforming Traditions in American Biology, 1880–1915*, Baltimore.

Mandelbaum, M. (1971), *History, Man and Reason: A Study in Nineteenth-Century Thought*, Baltimore.

Marx, K. (1967), *Capital*, vol 1, trans S. Moore and E. Aveling, New York.

* Mason, S. F. (1962), *A History of the Sciences*, New York.

Matthews, J. R. (1995), *Quantification and the Quest for Medical Certainty*, Princeton, NJ.

Maulitz, R. C. (1987), *Morbid Appearances: The Anatomy of Pathology in the Early Nineteenth Century*, Cambridge.

Maulitz, R. C. and Long, D. (1988), *Grand Rounds: One Hundred Years of Internal Medicine*, Philadelphia.

Mayr, E. (1982), *The Growth of Biological Thought: Diversity, Evolution and Inheritance*, Cambridge, MA.

Mayr, O. (1986), *Authority, Liberty and Automatic Machinery in Early Modern Europe*, Baltimore and London.

McClelland, C. E. (1980), *State, Society and University in Germany, 1700–1914*, Cambridge.

McIntosh, R. P. (1985), *The Background of Ecology: Concept and Theory*, Cambridge.

Meinel, C. (forthcoming), 'Modelling a Visual Language for Chemistry, 1860–1875', in S. de Chadarevian and N. Hopwood (eds), *Displaying the Third Dimension: Models in the Sciences, Technology and Medicine*, Stamford, CA.

Melhado, E. M. (1981), *Jacob Berzelius: The Emergence of his Chemical System*, Stockholm and Madison.

Melhado, E. M. (1992), 'Novelty and Tradition in the Chemistry of Berzelius (1803–1819)', in E. M. Melhado and T. Frängsmyr (eds), *Enlightenment Science in the Romantic Era: The Chemistry of Berzelius and its Cultural Setting*, Cambridge, 132–70.

Melhado, E. M. and Frängsmyr, T. (eds) (1992), *Enlightenment Science in the Romantic Era: The Chemistry of Berzelius and its Cultural Setting*, Cambridge.

Mendelsohn, E. (1997), 'Science, Scientists and the Military', in J. Krige and D. Pestre (eds), *Science in the Twentieth Century*, Amsterdam, 175–202.

** Merz, J. T. (1904–12), *A History of European Thought in the Nineteenth Century*, 4 vols, New York.

Metzger, H. (1918), *La Génèse de la Science des Cristaux*, Paris.

Miall, S. (1931), *A History of the British Chemical Industry*, London.

Midgley, M. (1995), *Beast and Man: The Roots of Human Nature*, rev. edn, London and New York.

Miller, D. (1991), *Material Culture and Mass Consumption*, Oxford and Cambridge, MA.

Miller, D. L.(1992), *Lewis Mumford: A Life*, Pittsburgh.

Miller, D. P. and Reill, P. H. (eds) (1996), *Visions of Empire: Voyages, Botany, and Representations of Nature*, Cambridge.

Mirowski, P. (1989), *More Heat than Light*, Cambridge.

Moore, J. R. (1979), *The Post-Darwinian Controversies: A Study of the Protestant Struggle to Come to Terms with Darwin in Great Britain and America, 1870–1900*, New York.

Morantz-Sanchez, R. M. (1985), *Sympathy and Science: Women Physicians in American Medicine*, New York and Oxford.

Morantz-Sanchez, R. M. (1992), 'Feminist Theory and Historical Practice: Rereading Elizabeth Blackwell', *History and Theory*, 31, 51–69.

Morrell, J. (1972), 'The Chemist Breeders: The Research Schools of Liebig and Thomas Thomson', *Ambix*, 19, 1–46.

Morrell, J. (1997a), *Science at Oxford, 1914–1939: Transforming an Arts University*, Oxford.

Morrell, J. (1997b), *Science, Culture and Politics in Britain, 1750–1870*, Aldershot.

Morrell, J. and Thackray, A. (1981), *Gentlemen of Science: Early Years of the British Association for the Advancement of Science*, Oxford.

Morris, R. J. (1976), *Cholera 1832*, London.

Morton, A. G. (1981), *History of Botanical Science*, London.

Moseley, R. (1978), 'The Origins and Early Years of the National Physical Laboratory: A Chapter in the Pre-History of British Science Policy', *Minerva*, 16, 222–50.

Mowery, D. C. and Rosenberg, N. (1989), *Technology and the Pursuit of Economic Growth*, Cambridge.

Mulkay, M. (1997), *The Embryo Research Debate: Science and the Politics of Reproduction*, Cambridge.

Mumford, L. (1934), *Technics and Civilisation*, New York.

Mumford, L. (1964), 'Authoritarian and Democratic Technics', *Technology and Culture*, 5, 1–8.

Musson, A. E. (1975), 'Joseph Whitworth and the Growth of Mass-Production Engineering', *Business History*, 17, 109–49.

Musson, A. E. and Robinson, E. (1969), *Science and Technology in the Industrial Revolution*, Manchester.

Needham, J. (1959), *A History of Embryology*, 2nd edn, Cambridge.

Nicolson, M. (1987), 'Alexander von Humboldt, Humboldtian Science, and the Origins of the Study of Vegetation', *History of Science*, 25, 167–94.

Nordenskiold, E. (1928), *History of Biology*, New York.

* North, J. (1994), *The Fontana History of Astronomy and Cosmology*, London.

Nye, M. J. (1996), *Before Big Science: The Pursuit of Modern Chemistry and Physics*, New York and London.

O'Brien, P., Griffiths, T. and Hunt, P. (1996), 'Technological Change during the First Industrial Revolution: The Paradigm Case of Textiles, 1688–1851', in R. Fox (ed.), *Technological Change*, Amsterdam, 155–76.

Olby, R. (1974), *The Path to the Double Helix: The Discovery of DNA*, London.

** Olby, R. C., Cantor, G. N., Christie, J. R. R. and Hodge, M. J. S. (eds) (1990). *Companion to the History of Modern Science*, London and New York.

Olesko, K. M. (1988), 'Commentary on Institutes, Investigations and Scientific Training', in W. Coleman and F. L. Holmes (eds), *The Investigative Enterprise*, Berkeley, CA, 295–332.

Olesko, K. M. (1991), *Physics as a Calling: Discipline and Practice in the Konigsberg Seminar for Physics*, New York and London.

Orel, V. (1984), *Mendel*, trans. S. Finn, Oxford.

Oudshoorn, N. (1994), *Beyond the Natural Body: An Archeology of Sex Hormones*, London.

Oudshoorn, N. (1998), 'Shifting Boundaries between Industry and Science: The Role of the WHO in Contraceptive R&D', in J.-P. Gaudillière and I. Löwy (eds), *The Invisible Industrialist*, Basingstoke, London and New York, 345–68.

Outhwaite, W. (1975), *Understanding Social Life: The Method Called Verstehen*, London.

Outram, D. (1984), *Georges Cuvier. Vocation, Science and Authority in Post-Revolutionary France*, Manchester.

Owens, L. (1997), 'Science in the United States', in J. Krige and D. Pestre (eds), *Science in the Twentieth Century*, Amsterdam, 821–37.

* Pacey, A. (1974), *The Maze of Ingenuity: Ideas and Idealism in the Development of Technology*, London.

Parascandola, J. (ed.) (1980), *The History of Antibiotics*, Madison.

Patterson, E. (1970), *John Dalton and the Atomic Theory*, New York.

Pauly, P. J. (1987), *Controlling Life: Jacques Loeb and the Engineering Ideal in Biology*, New York and Oxford.

Perez, J. F. and Tascon, I. G. (eds) (1991), *Description of the Royal Museum Machines*, Madrid.

Pérez-Ramos, A. (1988), *Francis Bacon's Idea of Science and the Maker's Knowledge Tradition*, London.

Pernick, M. (1985), *A Calculus of Suffering: Pain, Professionalism and Anesthesia in Nineteenth-Century America*, New York.

Peters, T. F. (1996), *Building the Nineteenth Century*, Cambridge, MA.

Phillips, I. A. (1989), 'Concepts and Methods in Animal Breeding 1770–1870', unpublished PhD thesis, UMIST.

Pick, D. (1993), *War Machine: The Rationalisation of Slaughter in the Modern Age*, New Haven, CT, and London.

Pickstone, J. V. (1973), 'Globules and Coagula: Concepts of Tissue Formation in the Early Nineteenth Century', *Journal of the History of Medicine*, 28, 336–56.

Pickstone, J. V. (1981), 'Bureaucracy, Liberalism and the Body in Post-Revolutionary France: Bichat's Physiology and the Paris School of Medicine', *History of Science*, 19, 115–42.

Pickstone, J. V. (1985), *Medicine and Industrial Society: A History of Hospital Development in Manchester and its Region, 1752–1946*, Manchester.

Pickstone, J. V. (1990), 'Physiology and Experimental Medicine', in R. C. Olby, G. N. Cantor, J. R. R. Christie and M. J. S. Hodge (eds), *Companion to the History of Modern Science*, London, 728–42.

Pickstone, J. V. (ed.) (1992a), *Medical Innovations in Historical Perspective*, Basingstoke and London.

Pickstone, J. V. (1992b), 'Dearth, Dirt and Fever Epidemics; Rewriting the History of British "Public Health", 1780–1850', in T. Ranger and P. Slack (eds), *Epidemics and Ideas: Essays on the Historical Perception of Pestilence*, Cambridge, 125–48.

Pickstone, J. V. (1993a), 'Ways of Knowing: Towards a Historical Sociology of Science, Technology and Medicine', *British Journal for the History of Science*, 26, 433–58.

Pickstone, J. V. (1993b), 'The Biographical and the Analytical: Towards a Historical Model of Science and Practice in Modern Medicine', in I. Löwy (ed.), *Medicine and Change: Historical and Sociological Studies of Medical Innovation*, Paris (Les Editions INSERM, John Libbey), 23–46.

Pickstone, J. V. (1994a), 'Museological Science? The Place of the Analytical/Comparative in Nineteenth-Century Science, Technology and Medicine', *History of Science*, 32, 111–38.

Pickstone, J. V. (1994b), 'Objects and Objectives in the History of Medicine', in G. Lawrence (ed.), *Technologies of Modern Medicine*, London, 13–24.

Pickstone, J. V. (1995), 'Past and Present Knowledges in the Practice of History of Science', *History of Science*, 33, 203–24.

Pickstone, J. V. (1996), 'Bodies, Fields and Factories: Technologies and Understandings in the Age of Revolutions', in R. Fox (ed.), *Technological Change: Methods and Themes in the History of Technology*, Amsterdam, 51–61.

Pickstone, J. V. (1997), 'Thinking over Wine and Blood: Craft-Products, Foucault, and the Reconstruction of Enlightenment Knowledges', *Social Analysis*, 41, 99–108.

Picon, A. (1992), *L'Invention de l'Ingénieur Moderne: L'Ecole des Ponts et Chaussées 1747–1851*, Paris.

Picon, A. (1996), 'Towards a History of Technological Thought', in R. Fox (ed.), *Technological Change*, Amsterdam, 37–49.

Pinell, P. (1992), *Naissance d'un Fléau: Histoire de la Lutte contre le Cancer en France (1890–1940)*, Paris.

Pinell, P. (2000), 'Cancer', in R. J. Cooter and J. V. Pickstone (eds), *Medicine in the Twentieth Century*, Amsterdam.

Pointon, M. (1993), *Hanging the Head: Portraiture and Social Formation in Eighteenth-Century England*, New Haven, CT, and London.

Pointon, M. (ed.) (1994), *Art Apart: Art Institutions and Ideology across England and North America*, Manchester.

Polanyi, M. (1958), *Personal Knowledge: Towards a Post-Critical Philosophy*, London.

Pomata, G. (1996), '"Observatio" ovvero "Historia": Note su Empirismo e Storia in Età Moderna', *Quaderni Storici*, 91, 173–98.

Porter, D. (ed.) (1994), *The History of Public Health and the Modern State*, Amsterdam.

Porter, D. (2000), 'The Healthy Body', in R. J. Cooter and J. V. Pickstone (eds), *Medicine in the Twentieth Century*, Amsterdam.

Porter, R. (1977), *The Making of Geology: Earth Science in Britain 1660–1815*, Cambridge.

Porter, R. (ed.) (1995), *Medicine in the Enlightenment*, Amsterdam and Atlanta, GA.

* Porter, R. (ed.) (1996), *The Cambridge Illustrated History of Medicine*, Cambridge.

* Porter, R. (1997), *The Greatest Benefit to Mankind*, London.

Porter, R. and Teich, N. (eds) (1992), *Scientific Revolution in National Context*, Cambridge.

Porter, T. M. (1990), 'Natural Science and Social Theory', in R. C. Olby, G. N. Cantor, J. R. R. Christie and M. J. S. Hodge (eds), *Companion to the History of Modern Science*, London, 1024–43.

Porter, T. M. (1995), *Trust in Numbers: The Pursuit of Objectivity in Science and Public Life*, Princeton, NJ.

Price, D. De S. (1965), *Little Science, Big Science*, New York.

Price, G. (1976), *The Politics of Planning and the Problems of Science Policy*, Manchester.

Proctor, R. N. (1991), *Value-Free Science? Purity and Power in Modern Knowledge*, London and Cambridge, MA.

Pumfrey, S., Rossi, P. L. and Slawinski, M. (eds) (1991), *Science, Culture and Popular Belief in Renaissance Europe*, Manchester and New York.

Raven, C. E. (1968), *English Naturalists from Neckham to Ray: A Study of the Making of the Modern World*, New York.

* Ravetz, J. R. (1971), *Scientific Knowledge and its Social Problems*, Oxford.

Ravetz, J. R. (1990), 'The Copernican Revolution', in R. C. Olby, G. N. Cantor, J. R. R. Christie and M. J. S. Hodge (eds), *Companion to the History of Modern Science*, London, 201–16.

Reader, W. J. (1970, 1975), *Imperial Chemical Industries*, vols 1 and 2, London.

Reddy, W. M. (1986), 'The Structure of a Cultural Crisis: Thinking about Cloth in France Before and After the Revolution', in A. Appadurai (ed.), *The Social Life of Things: Commodities in Cultural Perspective*, Cambridge, 261–84.

Reich, L. S. (1985), *The Making of American Industrial Research: Science and Business at G.E. and Bell, 1876–1926*, Cambridge, MA.

Reuleaux, F. (1876), *Kinematics of Machinery*, trans. A. B. W. Kennedy, London.

Rhodes, R. (1986), *The Making of the Atomic Bomb*, London and New York.

Ricardo, D. (1971), *On the Principles of Political Economy and Taxation*, ed. R. M. Hartwell, London (first published 1817).

Richards, R. J. (1987), *Darwin and the Emergence of Evolutionary Theories of Mind and Behavior*, Chicago.

* Richardson, R. (1988), *Death, Dissection and the Destitute*, London.

* Risse, G. B. (1999), *Mending Bodies, Saving Souls: A History of Hospitals*, New York and Oxford.

Ritvo, H. (1987), *The Animal Estate: The English and Other Creatures in the Victorian Age*, Cambridge, MA.

Roger, J. (1971), *Les Sciences de la Vie dans la Pensée Française de la XVIIIe Siècle*, Paris.

Roll, E. (1973), *A History of Economic Thought*, 4th edn, London.

Romer, A. (1970), '[Antoine-] Henri Becquerel (1852–1908)', *Dictionary of Scientific Biography*, vol. 1, New York, 558–61.

Rosenberg, C. E. (1976), *No Other Gods: On Science and American Social Thought*, Baltimore and London.

* Rosenberg, C. E. (1992a), *Explaining Epidemics and Other Studies in the History of Medicine*, Cambridge.

Rosenberg, C. E. (1992b), 'The Therapeutic Revolution: Medicine, Meaning and Social Change in Nineteenth-Century America', in C. E. Rosenberg, *Explaining Epidemics and Other Studies in the History of Medicine*, Cambridge, 9–31.

Rosenberg, C. E. (1992c), 'Florence Nightingale on Contagion: The Hospital as Moral Universe', in C. E. Rosenberg, *Explaining Epidemics and Other Studies in the History of Medicine*, Cambridge, 90–108.

* Rosenberg, C. E. and Golden, J. (eds) (1992), *Framing Disease: Studies in Cultural History*, New Brunswick, NJ.

Rosenberg, N. (1976), *Perspectives on Technology*, Cambridge.

Rosenberg, N. (1982), *Inside the Black Box: Technology and Economics*, Cambridge.

Rosenberg, N. and Vincenti, W. G. (1978), *The Britannia Tubular Bridge: The Generation and Diffusion of Technological Knowledge*, Boston.

Rossi, P. (1957), *Francis Bacon: From Magic to Science*, trans. S. Rabinovitch, Chicago.

Rossi, P. (1970), *Philosophy, Technology, and the Arts in the Early Modern Era*, New York.

* Rouse, J. (1987), *Knowledge and Power: Toward a Political Philosophy of Science*, Ithaca, NY, and London.

Rousseau, G. S. and Porter, R. (eds) (1980), *The Ferment of Knowledge: Studies in the Historiography of Eighteenth-Century Science*, Cambridge.

Rudwick, M. J. S. (1972), *The Meaning of Fossils: Episodes in the History of Paleontology*, New York.

Rudwick, M. J. S. (1976), 'The Emergence of a Visual Language for Geological Science, 1760–1840', *History of Science*, 14, 149–95.

Rudwick, M. J. S. (1980), 'Social Order and the Natural World', *History of Science*, 18, 269–85.

Rudwick, M. J. S. (1985), *The Great Devonian Controversy: The Shaping of Scientific Knowledge among Gentlemanly Specialists*, Chicago.

Rupke, N. A. (1994), *Richard Owen: Victorian Naturalist*, New Haven, CT.

Russell, C. A. (1996), *Edward Frankland: Chemistry, Controversy and Conspiracy in Victorian England*, Cambridge.

Russell, C. A. with Coley, N. G. and Roberts, G. K. (1977), *Chemists by Profession: The Origins of the Royal Institute of Chemistry*, Milton Keynes.

Russell, E. S. (1916), *Form and Function: A Contribution to the History of Animal Morphology*, London.

Sachs, J. von (1890), *History of Botany 1530–1860*, Oxford.

Salomon-Bayet, C. (1986), *Pasteur et la Révolution Pastorienne*, Paris.

Sanderson, M. (1972), *The Universities and British Industry, 1850–1970*, London.

Santillana, G. de (1959), 'The Role of Art in the Scientific Renaissance', in M. Clagett (ed.), *Critical Problems in the History of Science*, Madison and London.

Schabas, M. (1990), *A World Ruled by Number: William Stanley Jevons and the Rise of Mathematics*, Princeton, NJ, and London.

Schaffer, S. (1980a), 'Herschel in Bedlam: Natural History and Stellar Astronomy', *British Journal for the History of Science*, 13, 211–39.

Schaffer, S. (1980b), 'Natural Philosophy', in G. S. Rousseau and R. Porter (eds), *The Ferment of Knowledge: Studies in the Historiography of Eighteenth-Century Science*, Cambridge, 55–92.

Schaffer, S. (1992), 'Late Victorian Metrology and its Instrumentation: A Manufactory of Ohms', in R. Bud and S. E. Cozzens (eds), *Invisible Connections: Instruments, Institutions, and Science*, Bellingham, 23–56.

Schaffer, S. (1995), 'Where Experiments End: Tabletop Trials in

Victorian Astronomy', in J. Z. Buchwald (ed.), *Scientific Practice: Theories and Stories of Doing Physics*, Chicago and London, 257–99.

Schaffer, S. and Shapin, S. (1985), *Leviathan and the Air Pump*, Princeton, NJ.

Schneider, M. A. (1993), *Culture and Enchantment*, Chicago and London.

Schofield, R. E. (1969), *Mechanism and Materialism: British Natural Philosophy in an Age of Reason*, Princeton, NJ.

Schumpeter, J. (1976), *Capitalism and Social Democracy*, London (first published in 1942).

Schupbach, W. (1982), 'The Paradox of Rembrandt's "Anatomy of Dr. Tulp"', *Medical History*, supplement no. 2, London.

Schuster, A. (1900), *The Physical Laboratories of the University of Manchester*, Manchester.

Schuster, J. A. (1990), 'The Scientific Revolution', in R. C. Olby, G. N. Cantor, J. R. R. Christie and M. J. S. Hodge (eds), *Companion to the History of Modern Science*, London, 217–42.

Secord, A. (1994), 'Science in the Pub – Artisan Botanists in Early-Nineteenth-Century Lancashire', *History of Science*, 32, 269–315.

Secord, J. A. (1986), *Controversy in Victorian Geology: The Cambrian-Silurian Dispute*, Princeton, NJ.

Shaffer, E. S. (1990), 'Romantic Philosophy and the Organization of the Disciplines: The Founding of the Humboldt University of Berlin', in A. Cunningham and N. Jardine (eds), *Romanticism and the Sciences*, Cambridge and New York, 38–54.

Shapin, S. (1994), *A Social History of Truth: Civility and Science in Seventeenth Century England*, Chicago and London.

* Shapin, S. (1996), *The Scientific Revolution*, Chicago.

Shapin, S. and Schaffer, S. (1985), *Leviathan and the Air-Pump: Hobbes, Boyle, and the Experimental Life*, Princeton, NJ.

Sheal, J. (1976), *Nature in Trust: The History of Nature Conservation in Britain*, Glasgow.

Sheets-Pyenson, S. (1989), *Cathedrals of Science: The Development of Colonial Natural History Museums during the Late Nineteenth Century*, Montreal.

Shilts, R. (1987), *And the Band Played On: Politics, People and the AIDS Epidemic*, London.

Shinn, T. (1980), *L'Ecole Polytechnique 1794–1914: Savoir Scientifique et Pouvoir Social*, Paris.

Shinn, T. (1992), 'Science, Tocqueville, and the State: the Organisation of Knowledge in Modern France', *Social Research*, 59, 533–66.

Sibum, O. (1995), 'Reworking the Mechanical Values of Heat: Instruments of Precision and Gestures in Early Victorian England', *Studies in History of the Physicial Sciences*, 26, 73–106.

Simon, B. (1997), *In Search of a Grandfather: Henry Simon of Manchester, 1835–1899*, Leicester.

Sloan, P. R. (1990), 'Natural History, 1670–1802', in R. C. Olby, G. N. Cantor, J. R. R. Christie and M. J. S. Hodge (eds), *Companion to the History of Modern Science*, London, 295–313.

Smith, A. (1776), *The Wealth of Nations*, 1982 edn, Harmondsworth.

Smith, C. (1990), 'Energy', in R. C. Olby, G. N. Cantor, J. R. R. Christie and M. J. S. Hodge (eds), *Companion to the History of Modern Science*, London, 326–41.

* Smith, C. (1998), *The Science of Energy: A Cultural History of Energy Physics in Victorian Britain*, London

Smith, C. and Wise, N. M. (1989), *Energy and Empire: A Biographical Study of Lord Kelvin*, Cambridge.

Smith, J. G. (1979), *The Origins and Early Development of the Heavy Chemical Industry in France*, Oxford.

Smith, M. R. (ed.) (1985), *Military Enterprise and Technological Change*, London and Cambridge, MA.

* Smith, R. (1997), *The Fontana History of the Human Sciences*, London.

Spary, E. C. (1995), 'Political, Natural and Bodily Economies', in N. Jardine, J. A. Secord and E. C. Spary (eds), *The Cultures of Natural History*, Cambridge and New York, 178–96.

* Stacey, M. (1988), *The Sociology of Health and Healing*, London.

Stafleu, F. (1971), *Linnaeus and the Linnaeans: The Spreading of their Ideas in Systematic Botany*, Utrecht.

Stansfield, R. G. (1990), 'Could We Repeat It', in J. Roche (ed.), *Physicists Look Back: Studies in the History of Physics*, Bristol, 88–110.

Star, S. L. and Griesmer, J. R. (1989), 'Institutional Ecology, "Translations", and Boundary Objects: Amateurs and Professionals in Berkeley's Museum of Verterbrate Zoology, 1907–39', *Social Studies of Science*, 13, 205–28.

Stearn, W. T. (1981), *The Natural History Museum at South Kensington*, London.

Stemerding, D. (1991), *Plants, Animals and Formulae: Natural History in the Light of Latour's Science in Action and Foucault's The Order of Things*, Enschede.

Stewart, L. S. (1992), *The Rise of Public Science: Rhetoric, Technology and Natural Philosophy in Newtonian Britain*, Cambridge.

Stocking, G. Jr (ed.) (1985), *Objects and Others: Essays on Museums and Material Culture*, Madison.

Studer, K. E. and Chubin, D. E. (1980), *The Cancer Mission: Social Contexts of Biomedical Research*, Beverly Hills, CA, and London.

Sturdy, S. (1992a), 'From the Trenches to the Hospitals at Home: Physiologists, Clinicians and Oxygen Therapy', in J. V. Pickstone (ed.), *Medical Innovations in Historical Perspective*, Basingstoke and London, 104–23.

Sturdy, S. (1992b), 'The Political Economy of Scientific Medicine: Science, Education and the Transformation of Medical Practice in Sheffield, 1890–1922', *Medical History*, 36, 125–59.

Sturdy, S. (2000), 'The Industrial Body', in R. J. Cooter and J. V. Pickstone (eds), *Medicine in the Twentieth Century*, Amsterdam.

Sturdy, S. and Cooter, R. (1998), 'Science, Scientific Management and the Transformation of Medicine in Britain c1870–1950', *History of Science*, 36, 421–66.

Süsskind, C. (1973), 'Langmuir', *Dictionary of Scientific Biography*, vol. 8, New York, 22–5.

Sviedrys, R. (1970), 'The Rise of Physical Science at Victorian Cambridge', *Historical Studies in the Physical Sciences*, 2, 127–51.

Sviedrys, R. (1976), 'The Rise of Physics Laboratories in Britain', *Historical Studies in the Physical and Biological Sciences*, 7, 405–36.

Swann, J. P. (1988), *Academic Scientists and the Pharmaceutical Industry*, Baltimore.

* Taylor, C. (1989), *Sources of the Self*, Cambridge.

Temin, P. (1980), *Taking Your Medicine: Drug Regulation in the United States*, Cambridge, MA.

Temkin, O. (1973), *Galenism*, Ithaca, NJ.

Temkin, O. (1977), *The Double Face of Janus*, Baltimore.

Thackray, A. (1970), *Atoms and Powers: An Essay on Newtonian Matter-Theory and the Development of Chemistry*, Cambridge, MA.

Thackray, A. (1974), 'Natural Knowledge in Cultural Context: the Manchester Model', *American Historical Review*, 79, 672–709.

** Thackray, A. (ed.) (1992), *Science after Forty*, Osiris, 7.

* Thomas, K. (1971), *Religion and the Decline of Magic*, London.

* Thomas, K. (1983), *Man and the Natural World: Changing Attitudes in England, 1500–1800*, London.

Thomason, B. (1987), 'The New Botany in Britain, 1870–1914', unpublished PhD thesis, UMIST.

Thompson, E. P. (1978), 'The Peculiarities of the English', *Poverty and Theory*, London.

Thompson, S. P. (1898), *Michael Faraday: His Life and Work*, London, Paris, New York and Melbourne.

Timmermann, C. (2000), 'Constitutional Medicine, Neo-Romanticism and the Politics of Anti-Mechanism in Interwar Germany', *Bulletin for the History of Medicine*.

Travis, A. S. (1989), 'Science as Receptor of Technology: Paul Ehrlich and the Synthetic Dyestuffs Industry', *Science in Context*, 3, 383–408.

Travis, A. S. (1992), *The Rainbow Makers. The Origins of the Synthetics Dyestuffs Industry in Western Europe*, Bethlehem, PA, and London.

Travis, A. S., Hornix, W. J. and Bud, R. (eds) (1992), *Organic Chemistry and High Technology 1850–1950*, British Journal for the History of Science, special issue, 25:1.

Tuchman, A. (1988), 'From the Lecture to the Laboratory: The Institutionalization of Scientific Medicine at the University of Heidelberg', in W. Coleman and F. L. Holmes (eds), *The Investigative Enterprise*, Berkeley and Los Angeles, 65–99.

Tuchman, A. (1993), *Science, Medicine and the State in Germany: The Case of Baden, 1815–1871*, New York and Oxford.

Turner, F. (1980), 'Public Science in Britain, 1880–1919', *Isis*, 71, 589–608.

Turner, R. S. (1971), 'The Growth of Professorial Research in Prussia, 1818–1848: Causes and Context', *Historical Studies in the Physical Sciences*, 3, 137–82.

Turner, R. S. (1982), 'Justus Liebig versus Prussian Chemistry: Reflections on Early Institute-Building in Germany', *Historical Studies in the Physical Sciences*, 13, 129–62.

Turrill, W. B. (1959), *The Royal Botanic Gardens, Kew*, London.

Vess, D. M. (1974), *Medical Revolution in France 1789–1796*, Gainesville, FL.

Vincenti, W. (1990), *What Engineers Know and How they Know It: Analyticial Studies from Aeronautical History*, Baltimore.

Vogel, M. J. and Rosenberg, C. E. (eds) (1979), *The Therapeutic Revolution: Essays in the Social History of American Medicine*, Philadelphia.

Vos, R. (1991), *Drugs Looking for Diseases: Innovative Drug Research and the Development of Beta Blockers and the Calcium Antagonists*, Dordrecht.

Walsh, V. (1998), 'Industrial R&D and its Influence on the Organization and Management of the Production of Knowledge in the Public Sector', in J.-P. Gaudillière and I. Löwy, (eds), *The Invisible Industrialist*, Basingstoke, London and New York, 298–344.

Ward, W. R. (1972), *Religion and Society in England 1790–1850*, London.

Warner, J. H. (1985), 'The Selective Transport of Medical Knowledge; Antebellum American Physicians and Parisian Medical Therapeutics', *Bulletin of the History of Medicine*, 59, 213–31.

Warner, J. H. (1986), *The Therapeutic Perspective: Medical Practice, Knowledge and Identity in America, 1820–1885*, Cambridge, MA.

Warner, J. H. (1991), 'Ideals of Science and their Discontents in late Nineteenth Century American Medicine', *Isis*, 82, 454–78.

Warner, J. H. (1994), 'The History of Science and the Sciences of Medicine', in A. Thackray (ed.), *Critical Problems in the History of Science*, Osiris, 10, 164–93.

* Watson, J. D. (1968), *The Double Helix: A Personal Account of the Discovery of the Structure of DNA*, London.

** Weatherall, M. (1990), *In Search of a Cure: A History of Pharmaceutical Discovery*, Oxford.

Weber, M. (1949), '"Objectivity" in Social Science and Social Policy' (first published in 1904), in E. A. Shils and H. A. Finch (eds and trans), *Max Weber on The Methodology of the Social Sciences*, Glencoe, IL.

Webster, C. (1975), *The Great Instauration: Science, Medicine and Reform 1626–1660*, London.

Webster, C. (1982), *From Paracelsus to Newton. Magic and the Making of Modern Science*, Cambridge.

Weindling, P. (1992), 'From Medical Research to Clinical Practice: Serum Therapy for Diphtheria in the 1890s', in J. V. Pickstone (ed.), *Medical Innovations in Historical Perspective*, Basingstoke and London, 72–83.

Weiss, J. H. (1982), *The Making of Technological Man: The Social Origins of French Engineering Education*, Cambridge, MA.

Werskey, G. (1978), *The Visible College: A Collective Biography of British Scientists and Socialists of the 1930s*, London.

White, G. (1901), *The Natural History of Selborne*, ed. R. Mabey, Harmondsworth.

Whitley, R. (1984), *The Intellectual and Social Organisation of the Sciences*, Oxford.

Williams, L. P. (1971), 'Faraday', *Dictionary of Scientific Biography*, vol. 4, New York, 527–40.

Williams, L. P. (1987), *Michael Faraday: A Biography*, London and New York.

Williams, R. (1958), *Culture and Society 1780–1950*, London.

Williamson, G. S. and Pearse, H. I. (eds) (1938), *Biologists in Search of Material: An Interim Report on the Pioneer Health Centre, Peckham*, London.

Wilson, A. and Ashplant, T. G. (1988), 'Whig History and Present-Centred History', *Historical Journal*, 30, 1–16.

Wilson, D. (1983), *Rutherford Simple Genius*, Cambridge, MA.

Wise, G. (1985), *Willis R. Whitney, General Electric and the Origins of U.S. Industrial Research*, New York and Guildford.

Worboys, M. (1976), 'Science and British Colonial Imperialism 1895–1940', unpublished PhD thesis, 2 vols, University of Sussex.

Worboys, M. (1988), 'Manson, Ross and Colonial Medical Policy: Tropical Medicine in London and Liverpool, 1899–1914', in R. MacLeod and M. Lewis (eds), *Disease, Medicine and Empire: Perspectives on Western Medicine and the Experience of European Expansion*, London, 21–37.

Worboys, M. (1992), 'Vaccine Therapy and Laboratory Medicine in Edwardian Britain', in J. V. Pickstone (ed.), *Medical Innovations in Historical Perspective*, Basingstoke and London, 84–103.

* Worboys, M. (2000), *Spreading Germs: Disease Theories in Medical Practice in Britain 1865–1900*, Cambridge.

Worster, D. (1985), *Rivers of Empire: Water, Aridity and the Growth of the American West*, New York.

Worster, D. (1994), *Nature's Economy: A History of Ecological Ideas*, 2nd edn, Cambridge.

Young, R. M. (1985), *Darwin's Metaphor: Nature's Place in Victorian Culture*, Cambridge.

Yoxen, E. (1983), *The Gene Business: Who Should Control Biotechnology?* London and Sydney.

Index